Electronic Engine Management

Reference Manual

Frank 'Choco' Munday

Published in 2004 by
Graffiti Publications Pty Ltd
69 Forest Street, Castlemaine, Victoria, Australia
Phone international 61 3 5472 3653
Fax International 61 3 5472 3805
Email: graffiti@netcon.net.au
Website: www.graffitipub.com.au

Copyright 2004 by Frank "Choco" Munday
Publisher: Larry O'Toole
Design: Michael deWolfe
Production: Mary-Anna Brennand

Printed by Centre State Printing, Maryborough, Victoria

All rights reserved. With the exception of quoting brief passages for the purposes of review, no part of this publication may be reproduced without prior written permission of the publisher.

Graffiti Publications books are also available at discounts in bulk quantity for industrial or sales promotional use. For details contact Graffiti Publications PH. 61 3 5472 3653.

Printed and bound in Australia

ISBN 0 949398 90 X

table of contents

	Table of Contents	**3**
	List Of Figures	**6**
Chapter 1	**Introduction**	**11**
	Hot Rods in the New Millennium	11
	Balancing Environment with Horsepower	11
	Electronic Fuel Injection (EFI)	12
	Disclaimer	14
	EFI Safety	15
Chapter 2	**An Overview of Engine Management Systems**	**19**
	Introduction	20
	Why Not Stick with Points and Carbs?	20
	What Is EFI?	20
	Fuel Delivery	22
	• Throttle Body Injection	22
	• Multi-Point Injection	24
	• Batch vs Sequential Injection	25
	Putting EFI and EMS to Work	26
Chapter 3	**Controlling Fuel**	**31**
	The Basics	31
	Introduction	32
	• Open Loop and Closed Loop	32
	• Sensors and Actuators	33
	Fuel Control	34
	Engine Control at Idle, Cold Starts, Cruise and WOT	35
	Controlling Nitrous Oxide	37
	How Does Nitrous Oxide Work?	37
	Why NO2?	38
	Nitrous Management	38
Chapter 4	**Controlling Ignition**	**41**
	The Basics	41
	Introduction	42
	Electronic Ignition Management	42
	• Distributor Control	42
	• Crankshaft Trigger Control	45
	• Distributorless Ignition (Coil Packs)	46
	Converting Electronic Ignition for use with Delco/Kalmaker ECM	49
	• Magnetic Distributor	49
	• Hall Effect Distributor	50
	• Kalmaker and Distributorless Ignition Systems	50
Chapter 5	**Sensors and Actuators**	**53**
	• Sensors, Actuators	53
	• Fuel Pump Relay	63
	• Fuel Pressure Regulator	63
	Fuel Injectors	65
	• Injector Types	65
	• Injector Choice	66
	• Fuel Transfer and Sealing	67
	• Flow Rates / Pressure	67

table of contents

Chapter 6	**The Engine Control Module**		**71**
	• Which System?		71
	• Programming your EMS for Engine Modifications		72
	• Choosing an Engine Management System		73
	Kalmaker		75
	• The Software		77
	• Using the Delco ECM and Kalmaker		78
	• Street Pro 3		78
	• Tuning the Engine Management System		79
Chapter 7	**Wiring and Auto Electronics**		**83**
	The Alternator		83
	Alternator Rating		85
	Starter Motors		87
	Wiring, RFI and EMI		88
	Wiring Harness		90
	Relays		91
	Tachometers		92
Chapter 8	**Converting to EFI**		**95**
	Convert to EFI		95
	Conversion Alternatives		97
	Choosing a Manifold		98
	Converting a Carb Manifold		99
	Adapting Throttle Bodies		104
Chapter 9	**Appendix**		**109**
	Fuel Rail Extrusion		109
	O2 sensor bungs		110
	Mopar Performance Electronic Ignitions		110
	Calculations, Estimates and Equations		111
	Throttle Bodies and Adapters		111
	Remote IAC Valves		112
	EFI and EMS Businesses and Groups		112
	Kalmaker		113
	GM Engine Management System ECM usage		115
	Further Reading		**115**
	Wire Gauge Tables		**115**
	Oxygen Sensor Wire Colours		**117**
	Fuel Injector Data Table		**118**
	Injector Flow Rate		**122**
	List of Abbreviations		**123**
	Alphabetical Index		**127**

list of figures

FIGURE 1.	A Dinosaur	13
FIGURE 2.	Reverse Current Protected Jumper Leads	15
FIGURE 3.	Fuel System Disassembly	16
FIGURE 4.	The Engine Management System Computer	19
FIGURE 5.	Basic Fuel Injection Diagram	21
FIGURE 6.	Throttle Body Injection	23
FIGURE 7.	GM Sequential Fuel Injection V8 (Holden)	24
FIGURE 8.	Megasquirt Computer	26
FIGURE 9.	Stack Injector	27
FIGURE 10.	Mike Davidson's (Flat Attack Racing) Hilborn style flathead EFI conversion	28
FIGURE 11.	Big Block Chevy Stack Injection	29
FIGURE 12.	Location of the Oxygen Sensor	32
FIGURE 13.	Mass Airflow Meters	34
FIGURE 14.	Base Idle Screw on a typical GM Throttle Body, as viewed from underneath.	36
FIGURE 15.	Nitrous Oxide Injection	37
FIGURE 16.	Electronic Distributor	43
FIGURE 17.	MSD 6AL Trigger Box	44
FIGURE 18.	Wasted Spark Ignition System	47
FIGURE 19.	DFI Ignition System	48
FIGURE 20.	MAT and connector. Image courtesy of MSD.	53
FIGURE 21.	CTS. Photo courtesy of MSD.	54
FIGURE 22.	Location of CTS.	54
FIGURE 23.	Heated Oxygen Sensor. Photo courtesy of MSD	55
FIGURE 24.	O2 Sensor bung.	56
FIGURE 25.	Oxygen Sensor and weld in bung. Heated O2 Sensor shown.	56
FIGURE 26.	MAP Sensor.	57
FIGURE 27.	Throttle Position Sensor (typical GM)	58
FIGURE 28.	TPS in place.	59
FIGURE 29.	Idle Air Control valve.	59
FIGURE 30.	Remote IAC Valve	60
FIGURE 31.	Mass Air Flow Sensor. GM LT1 Shown.	62
FIGURE 32.	IAC Valves.	62
FIGURE 33.	Fuel Pressure Regulator. Image courtesy of Accel.	63
FIGURE 34.	Adjustable Fuel Pressure Regulator.	64
FIGURE 35.	Fuel Injectors	66
FIGURE 36.	Securing the Injectors.	67
FIGURE 37.	Securing Fuel Rails	68
FIGURE 38.	EPROM	72
FIGURE 39.	GM Delco ECM	77

FIGURE 40.	High Output Alternators	84
FIGURE 41.	Hi Torque Starter Motor	88
FIGURE 42	Computer Mounting	89
FIGURE 43.	Relay and Relay Panel	91
FIGURE 44.	Relay	91
FIGURE 45.	GM TPI	95
FIGURE 46.	Ford 5.0L Custom Plenum by Marshall Perron	96
FIGURE 47.	Toyota Quad Cam V8	96
FIGURE 48.	Stack Injection Alternative.	98
FIGURE 49.	Holley carb bases used as throttle bodies on this converted tunnel ram.	104
FIGURE 50.	Providing Air Flow with Carb Spacers	104
FIGURE 51.	Three Bolt TBI Throttle Body Adapter	105
FIGURE 52.	Billet 4 bbl Throttle Body from EFI Hardware	106
FIGURE 53.	Twin Tech adapter	106
FIGURE 54.	Fuel Rail Extrusion	109

about the author

I first started fiddling around with old cars when I was about twelve years old. I liked the concept of an old car that ran and performed like a new car, and it became an obsession that is still with me today. I joined the Royal Australian Navy at fifteen and served for twenty years as an Electronic Technician, travelling the world and even living in the US for a few years when commissioning a new ship. I bought my first Hot Rod, a 1927 T roadster, in 1975 and built a chopped, channelled and flamed 1936 Ford Coupe several years later that was to be my ride for the next 20 years. My current project is a 1936 Plymouth Coupe with an EFI converted 360 Chrysler V8 engine.

I always had a flair for writing, and, after leaving the Navy in 1990, I became a Technical Author for a Defence Electronics company in Canberra, in the Australian Capital Territory, where I have lived since 1977. I worked with an ex-Navy colleague who owned a 1986 Corvette, and he provided me with a factory engine manual (as well as the use of his car) so that I could help a Hot Rodding friend reproduce the factory wiring of a small block Chevy TPI in his '32 Ford Tudor. When I noticed some discrepancies between the wiring in the manual, the wiring in the Corvette and the information provided with the cut out 350 TPI engine, I re-wrote much of the information to make it easier to follow. This became the draft of my first book, the Small Block Chevy Tuned Port Fuel Injection System. A mutual friend showed a copy of what I had written to Larry O'Toole, publisher, who convinced me to continue developing the book until it was a finished product.

This first exposure to the mysteries of Electronic Fuel Injection and the success of this first book convinced me to use my skills as a writer, an electronic technician, a computer professional and a hot rodder to bring modern engine management systems into the modified car enthusiasts world. My second book, the Custom Auto Electronics Reference Manual, was an attempt to describe auto electronics in a simple, but hands-on way, with data and information that could be put to good use.

During all my research, I was constantly frustrated by the extremely high technical level of the information that I was given as references. Nowhere could I find simple, basic instructions or entry-level data, so I had to create it myself. In doing so, I built the systems that I wrote about, as this is the only way I know to achieve the kind of intimacy between the practical application and the written description. For this reason, writing this book was a most laborious task that turned into a monster! I realised early in the process that I had bitten off more than I could chew, but I was determined to finish it, so I just chewed harder!

I had a lot of help from many highly skilled and experienced technicians in the fields of auto electronics and electronic fuel injection, many of whom I have only met over the phone or the internet. While they all know who they are, I would like to single out a few names for special mention, because, without them, this book would still be an idea and an outline on my computer.

- Ken Young. The genius behind Kalmaker, Ken obviously invested thousands of hours and as many dollars in the development of his software. Kalmaker won't make him a millionaire, but it should!
- Alan Gibbs. If there's one person who could be crowned the "Hands On King" of EFI conversions and modifications, it's Alan Gibbs from Injection Connection in Perth, Western Australia. A real busy guy, Alan contributed more to this book than any single entity, even me! I would like to thank Alan for his valuable contribution.
- Tex Smith. To be asked by Tex Smith to contribute to the Ron Ceridono book, "The Complete Chrysler Hemi Engine Manual", was an honour. To see it in print was a shock! I would like to state here that Tex is an inspiring guy, who took one look at my 1936 Plymouth Coupe EFI project and said "It's got a roof. It ain't a Hot Rod". He also said "You write good"! Thanks Tex.
- Dennis Overholser, Painless Wiring. The staff and technicians at Painless Wiring have all been supporters of the Hot Rod Handbooks publications, and Dennis has bent over backwards to accommodate me with hardware and technical data when compiling these books. A more professional company does not exist.
- Autotronic Controls Corporation, MSD Ignition Systems. I have used MSD ignition systems on all of my projects, and the MSD team in El Paso, Texas, were always ready and willing to share their knowledge and contribute to the success of Hot Rod Handbooks. Many thanks for your support.
- Larry O'Toole. Besides being the Publisher, Larry is also a dyed-in-the-wool Hot Rodder who knows exactly what to say and how to say it. Thanks for letting me know that I'm not wasting my time!

This book is dedicated to the Hot Rodders of the world, who all share a common bond that is unknown to other "normal" people. Don't ask us what that bond is, it just "is". The love of old cars that go like the wind is a phenomena that just won't die, and with the coming of age of Electronic Fuel Injection, a new page in the hobby has been turned. All I ask of my fellow Hot Rodders is that you take a peek and maybe consider tossing that carburettor once and for all. You won't regret it!

Choco Munday. Canberra, 2004.

1 introduction

Welcome to another production of Hot Rod Handbooks.

This book is an attempt to bring to modified car enthusiasts information and guidance on modern, efficient, auto electronic and electrical systems that will work well in our cars. It expands on the previous books to provide a practical guide to converting, installing and maintaining the following systems:

- Electronic Fuel Injection Systems
- Electronic Ignition Systems
- Engine Management Systems
- New, compact, high output alternators
- Modern wiring systems and kits

This book examines the practical application of these systems and covers most of the available technologies. The author acknowledges that there are many players in the aftermarket field, but only the stand out leaders are mentioned.

Hot Rods in the New Millennium

It's no secret that today's cars are doing more with less. Gone are the muscle cars of the late sixties and early seventies, but the good news is that, in limited numbers, we are returning to the days of one horsepower per cubic inch, all on 90 octane fuel! However, one look inside the engine compartment of a 21st Century V8 powered car, and the average hot rodder goes dizzy trying to find a way through the maze of hoses, wires and pipes. With many, it's back to the old carb/points setup for simplicity's sake - a waste of technology! The fact is, there are more EFI engines around now than the old carb/points engines. The prices are good, the blocks themselves are the same, so there's no reason why we can't make them fit. Then there's some of the more interesting new power plants, such as Dodge/Chrysler's Viper V10, the resurrection of the legendary Hemi, the Ford SOHC 5.0L and the DOHC Cobra 5.0L V8, the Toyota quad cam V8 and even the humble small block chevy has a V12 cousin on the market, all controlled and monitored by a computer.

Balancing Environment with Horsepower

First of all, let's just clear the air (no pun intended) of a couple of issues. I do not subscribe to the theories perpetrated by oil companies and (well meaning) environmentalists, that our modified cars are gross pollutants. A well tuned, well maintained V8 carburetted engine will emit as much pollution (in fact, probably less) as an average five year old car.

That's because catalytic converters have a life span of about 4 - 5 years - after that, they allow some pretty gross stuff to leak from our exhausts, far nastier than the lead that governments had us believe would kill us all![1] We also know that the "ozone holes" scare of a few years ago is a scam, and that what was happening over the poles was a naturally occurring phenomenon[2]. Having said that, there is still a lot of junk in the atmosphere that is gradually being reduced through the application of sensible, realistic and scientifically endorsed methods – like electronic fuel injection.

Electronic fuel injection. (EFI).

So why are modern cars being built this way? Well, to start with, the seventies was the era that brought us the anti-pollution movement. World governments introduced tough new laws to reduce the threat of pollution from the hundreds of thousands of cars manufactured each year. This meant more efficient combustion methods had to be invented. No way can a carburettor and points system efficiently reduce pollution. Fuel Injection and Electronic Ignition were the only avenues for the car manufacturers to take in order to make their products compliant with the new laws. Engine designers were no longer allowed to focus on performance above all else. Engines had to be clean and efficient, so performance had to suffer, and suffer it did. But with today's technology, we can now squeeze 300HP from a small block chevy and you can blow-dry your hair with the exhaust gas! What's even better is the car manufacturers spent millions getting all this right so we could have:

- Better fuel economy - Fuel economy has doubled in the same size engines over the last twenty years or so.

- More torque and power - Intake manifolds perform at their best with a dry flow.

- Improved throttle response and drivability - No hesitation or off-idle stumble when cold, no vapour lock when hot.

- Engine life increases - Decreased oil dilution results in less engine wear.

Factory, computer controlled engines can be modified. They can be supercharged, turbocharged, nitrous injected or just rebuilt with bigger cams, better balancing, more efficient heads - just the way we've always done it. But there is a downside.

---

1. "Hushed Up Dangers", Explore, Vol 5, No 5/6, 1994. S. Grose, "Choose Your Poison", Canberra Times, March 26, 1994. "Scientists Debate Carcinogenic Risk of Cars", Australian, August 13, 1994. G. Allum, "Is There an Old Car in Your Future?", Restored Cars, No 104/1993. D. Maddock, "Leaded V Unleaded Petrols", Australian Health & Healing, Vol 14/1. C. Simons, "The Lies of Unleaded Petrol", Nexus, April/May, June/July 1995. P. Sawyer, Green Hoax Effect, Groupacumen Publications, Victoria, 1990. http://www.greenleft.org.au/back/1995/198/198p11.htm

2. http://www.despatch.cth.com.au/Books_D/environ1.htmP. Sawyer, Green Hoax Effect, Groupacumen Publications, Victoria, 1990

- Engine peripherals - Single wire alternators that pump out sufficient electricity to handle huge stereo systems, electrically operated water pumps, oil pumps, power steering, fuel injection and other devices.

- Wiring systems - Wiring MUST be of the highest quality.

- Computers - The ECM needs to be shielded from power surges and spikes, the grounding must be perfect and each connection must offer the least electrical resistance possible.

- Knowledge Overflow - There's just so much you need to know when it comes to putting it all together.

Hot Rod Handbooks shall attempt to remove the mystery and give the average hot rodder the right amount of information required to build an EFI engine and wire it up in a non-factory environment.

This book is NOT about improving the performance of a late model car by fiddling with the engine management system, although readers will be capable of doing so with the aid of the information provided in these pages. Rather, it is about converting/updating from older carb/points engines to a more efficient version using today's technology.

By the time you finish this book, you will be too embarrassed to look inside the engine bay of your Hot Rod, Custom, 4WD, or Street Machine because it's got a carb. The only reason to have such a dinosaur system is when we decide to "restore" an old Hot Rod (the nostalgia look) but even in these areas we can use a bit of modern technology.

FIGURE 1. A Dinosaur
Even carbs as good looking as these are way too inefficient and difficult to maintain, given the alternative of EFI. Today's technology can be utilised on all kinds of hot rod projects and still maintain that traditional look and feel. It just requires a level of hot rod ingenuity.

Disclaimer

Hot Rod Handbooks is dedicated to providing accurate and useful information, and this is the case to our knowledge. All recommendations are made without guarantee on the part of the author, the publisher or their agents, and they disclaim any liability incurred in connection with the use of the information in this book.

Many words, models, names, designations and identifications in this book are the sole property of the trademark holder and are used for identification purposes only. This is not an official publication.

EFI Safety

- Electronic equipment is sensitive to current surges and voltage spikes. The most common cause is connecting components while battery power is on and when using an arc/mig welder. **Disconnect the battery before removing and replacing EFI components. Disconnect the ECM and the battery when welding on a vehicle.**

- Reverse current *will* damage electronic equipment. Ensure the battery is connected the correct way, negative (-) to the body/chassis.

- Wrong voltages can damage equipment. The ECM and powered sensors operate internally on 4.5 volts. Only connect 12 Volt power to the power feed wires of the ECM.

- Disconnect the battery before charging and when removing/replacing any of the EFI components.

- Use *reverse-current protected* jumper leads when jump starting your EFI vehicle. Jump starts should only be a last resort, in an emergency. NEVER use normal, unprotected jumper leads!

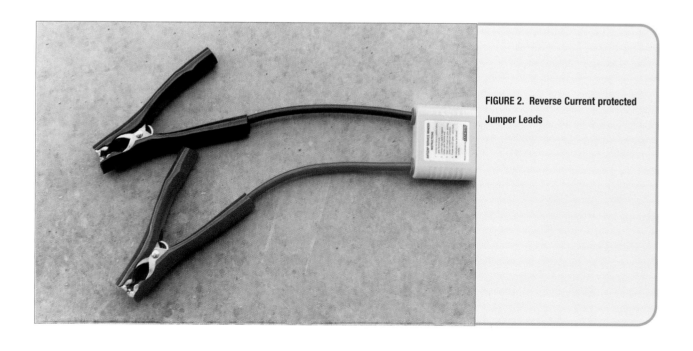

FIGURE 2. Reverse Current protected Jumper Leads

Use only THIS PROCEDURE when jump starting your car:

1. Connect the Red cable clamp to the positive post of the dead battery.

2. Connect the clamp at the other end of the Red cable to the positive post of the good battery.

3. Connect the Black cable clamp to the negative post of the good battery.

4. MAKE THE FINAL CONNECTION ON THE ENGINE BLOCK OF THE STALLED CAR - as far away as practicable from battery. *Make certain the cars are not touching.*

5. Attempt to start the stalled car with good engine OFF. If you can't get the stalled car to kick over after about 15 seconds, stop and check the stalled car again.

6. Remove the cables using the exact reverse procedure.

- **Do not** test for the presence of 12 Volt power by touching wires together and looking for a spark.

- Prolonged exposure to heat can adversely affect electronic equipment. Remove the ECM before placing the car in a paint oven.

- Fuel from an EFI system is under high pressure, and stays under pressure even when the engine has stopped. Take care with fuel lines, as fuel could spray into your eyes if a fuel line or hose is being disconnected.

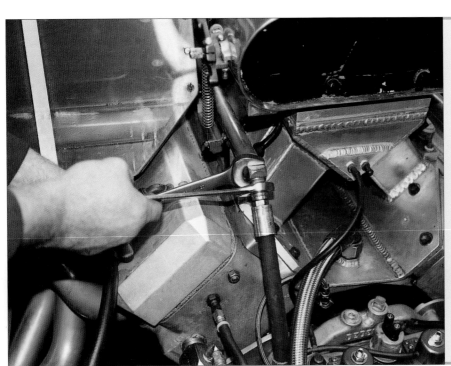

FIGURE 3. Fuel System Disassembly
The fuel system is under high pressure, take care with fuel lines as fuel could spray into your eyes when a fuel line is disconnected.

- To work on any part of the fuel system, be sure to follow this procedure to bleed fuel pressure off safely:

1. Remove the Fuel Pump fuse.
2. Start the engine and let it run until it stalls from lack of fuel.
3. Crank the engine again for a few seconds.
4. Turn the ignition OFF.
5. Let the engine cool.
6. It is now safe to work on the fuel system.

- Electronic Ignitions use MUCH HIGHER voltages than your old points system. Treat trigger boxes with respect and always use at least 8mm good quality carbon spiral core ignition cables.
  NEVER USE SOLID CORE WIRE CABLE ON ELECTRONIC IGNITION SYSTEMS!

2 an overview of engine management systems

An Engine Management System (EMS) is one which controls the fuel and ignition delivery for you. In the old days, an EMS was the accelerator, choke, float, jets, power valves, curb idle, air idle screws (for fuel management), points gap, centrifugal advance weights, vacuum advance and how far you twisted the distributor around (for ignition management). Of course, there was a whole lot more to consider, too, but you get the picture. Today's engine management systems are fully automated by a small computer we shall call the Electronic Control Module (ECM). It's also called the Electronic Control Unit (ECU), but for our purposes, they are one and the same thing.

FIGURE 4. The Engine Management System Computer
Also known as the Engine Control Module (ECM), the Electronic Control Unit (ECU), the Powertrain Control Module (PCM) and other nerdy type acronyms. We'll try to stick to either "the computer" or just ECM.

Overview of EMS

Introduction

Fuel delivery, fuel control, spark delivery and spark control are all managed for optimum performance automatically and dynamically - in other words, it's managed all the time the engine is running.

Why Not Stick with Points and Carbs?

Why not, indeed. If it's not broken, you can't fix it, right? Well, this may be true for many things but not when it comes to installing a non-factory drivetrain in your vehicle. The engine may be easy to work on with the familiar carb and the comfy, although dodgy, distributor containing one or two sets of points. Now let's be honest - wouldn't it be great to get a four barrel carb straight out of the box that maintained a constant, perfect air:fuel ratio, no matter what the engine speed or load? Yeah, and PIGS MIGHT FLY! Getting a carb in a high performance application to run right needs an expert with lots of time to spare. Then there's the manifold selection - which one is right for you? Now try to select a manifold that delivers equal amounts of air/fuel to each slug at the same velocity. Ahh, here come the pigs again! Now, what about a set of points that fires the spark plugs just at the right angle of the crankshaft's rotation? Regardless of engine speed, these magical wonder-points also fire the same amount of energy regardless of engine speed or load. If you believe this is possible with this old technology, it's time for a reality check.

What Is EFI?

Fuel Injection itself is not new. It has been used on street driven cars since the thirties. There are plenty of historical books around on the subject, so we won't dwell on that here. However, let's just remind ourselves of the basics of fuel injection:

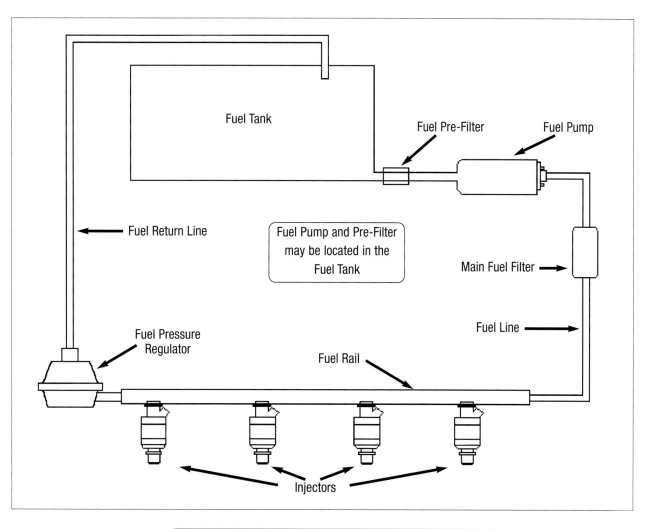

FIGURE 5. Basic Fuel Injection Diagram

Fuel Tank.	It's just a fuel tank. It should be baffled to prevent fuel sloshing around which can cause fuel starvation.
Fuel Filter.	A Fuel Injection fuel pump is a positive displacement pump. Contamination from dirty fuel can stall, and eventually destroy, the pump. A *pre-filter* prevents foreign matter from entering the fuel pump and a good quality canister type fuel filter should be used after the fuel pump.
Fuel Pump.	A high-pressure pump (usually 45 - 60 psi) supplies fuel to the injectors.
Fuel Line.	Transports the fuel from the pump to the fuel rail. Due to the high pressure involved, steel, aluminium and/or high pressure flexible hose are required.
Fuel Rail.	A small fuel manifold which distributes fuel to the injectors in a straight line. Because fuel injectors *pulse* (open and close) a fuel rail is required to contain the pulsing fuel.
Injectors.	Electrically operated valves which, when open, allow fuel to be injected into the engine under high pressure. Fuel Injectors are connected to the fuel rail via a clip and 'O' ring which has to contain the high pressure within the fuel system.
Fuel Pressure Regulator.	Maintains a constant pressure to the injectors, depending on injector size and engine demand. It also returns excess fuel to the Fuel Tank
Fuel Return Line.	Bleeds excess fuel back to the fuel tank.

Fuel Delivery

EFI is the combination of Fuel Injection and a computer. The basic engine is the same, it's just that the fuel is squirted into the engine by an electrically operated Injector, as opposed to being sucked in through a carburettor, in one of two broadly described methods.

- Throttle Body Injection (TBI)

- Multi-Point Injection (MPI)

Throttle Body Injection

The first mass produced EFI engines were of this design, and were virtually a replacement of the carb. Put simply, it is a *Wet Manifold* method of fuel injection. The *Throttle Body* (in much the same manner as a carburettor) lets air into the intake manifold and the Injectors squirt the fuel into the manifold at the base of the throttle body. The air and fuel combine and the mixture is distributed to the cylinders via the intake runners in the same manner as a carburettor. Also called *Central Fuel Injection* (CFI), this form of fuel injection is not quite as efficient as *Multi Point Injection*. The wet manifold design also has a disadvantage because of the nature of the fuel to condense, pool, or come out of suspension in the manifold during hard cornering, but TBI is still far better than a carb, as the amount of fuel and air is constantly maintained at an optimum ratio across the rev range.

It should be noted that TBI is NOT a "smart carburetor". That would give carbs a level of intelligence they don't deserve. TBI shares the wet manifold design but is an efficient, computer controlled fuel injection system, differing from TPI only in how the fuel is introduced to the cylinders. A carb relies on highly inaccurate and often inconsistent engine generated vacuum to feed and mix the fuel, so TBI still has the advantages of driveability, economy, efficiency, power, reliability, and throttle response.

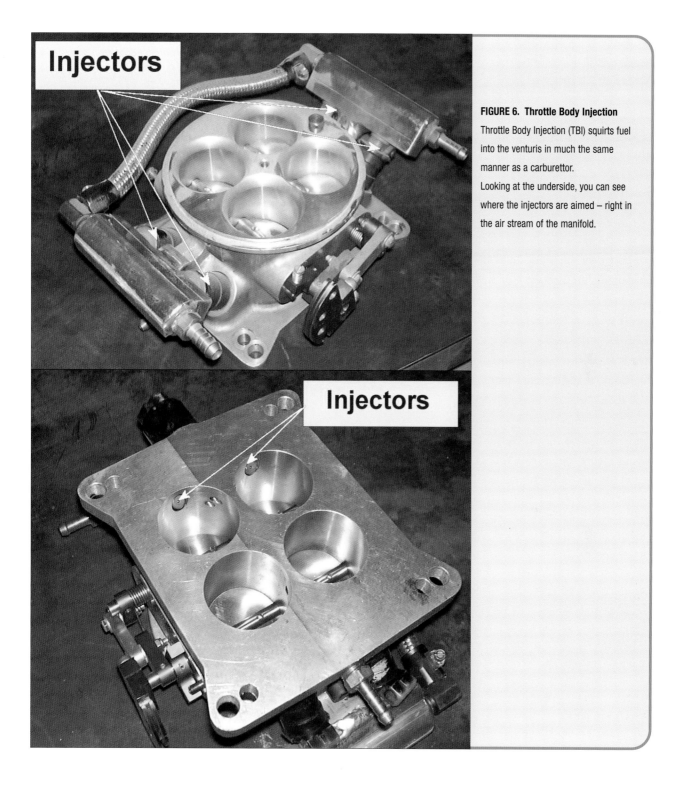

FIGURE 6. Throttle Body Injection
Throttle Body Injection (TBI) squirts fuel into the venturis in much the same manner as a carburettor.
Looking at the underside, you can see where the injectors are aimed – right in the air stream of the manifold.

Multi-Point Injection.

This method (also known as Multi-Port Injection) squirts the fuel directly at the base of each intake valve. The Throttle Body carries out the same function as TBI (regulates the air that enters the engine) but it's just the air that travels through the intake runners, and is, therefore, called a Dry Manifold system. This is arguably the better of the two EFI systems. It is more complex and more expensive than TBI, but the advantages will be evident as you gain an insight to the way it works. For now, remember this:

Dry Inlet Manifold. Higher inlet flow rates are achievable with a dry inlet manifold. The fuel and air don't mix together on their way through the manifold - this occurs right at the very end. The downside is that there is no wet flow mixture of air and fuel to physically cool the intake. This has the effect of reducing the temperature of the intake charge, which increases performance. Because heated air is passing through the dry manifold, the manifold gets hot, which heats the incoming air to the cylinders. Air is not a good thermal conductor, so the dry intake manifold has a tendency to retain heat over longer periods of time than a wet flow manifold design.

Equal Distribution of Fuel. The Air/Fuel mixture to each cylinder is consistent. This also means higher compression ratios are available (just remember that gasoline from the pump may be unsuitable for anything over 10:1 compression).

Accurate Distribution of Fuel. At any given RPM and engine load, every cylinder gets the optimum amount of air and fuel which will squeeze out the highest horsepower.

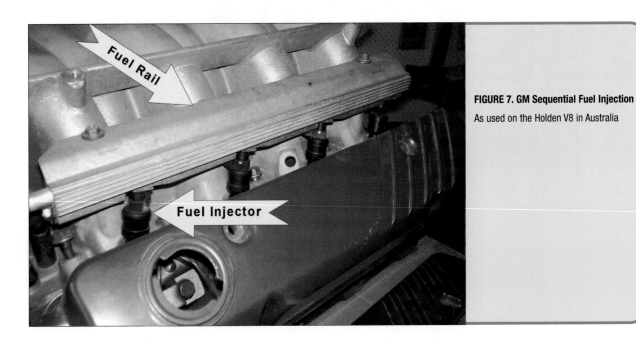

FIGURE 7. GM Sequential Fuel Injection
As used on the Holden V8 in Australia

Multi Point Injection (MPI) systems deliver the fuel in one of two ways:

- Batch Injection

- Sequential Injection

Batch Injection

An injection of fuel takes place simultaneously for each valve during every crankshaft revolution. During the first of two revolutions that complete a firing sequence, fuel is injected at the base of the intake valve while it is still closed. The second injection is sprayed into the airstream entering the chamber while the valve is open.

A modified version of Batch Injection is Bank Injection, where the cylinders are divided into Banks, dependant on the firing order. On the first revolution of the crankshaft, one bank of cylinders receives an injection of fuel. On the second revolution, another bank gets its turn, and so on.

Sequential Injection

A more sophisticated version, the injector squirts fuel ONLY when the intake valve is opening. The Engine Management System knows the relative position of each cylinder during the engine's cycle (by a signal from the Camshaft Position Sensor) then it can fire the injectors at the optimum time for that cylinder.

Batch vs Sequential Injection

There are advantages and disadvantages with both these systems. Batch Injection, on the face of it, uses more fuel, but makes use of it because fuel is already present and in vapour form at the intake stroke. It also requires far less sophisticated electronics, as the ECM only needs to provide two injection events per cycle, while Sequential Fuel Injection (SFI) requires eight (for a V8). On the other hand, SFI has the advantage of very fine tuning and fast throttle response. On a standard engine, SFI delivers slightly better torque figures at low to mid range engine speeds, but Batch Injection systems have proven very satisfactory and have the advantage of simplicity. A SFI system needs a Crankshaft Position Sensor to trigger injection sequences. Batch Injection only needs a pulse from the ignition system such as a tachometer output pulse from an MSD box, a magnetic pulse from an electronic distributor or a pulse from a crankshaft trigger wheel to time fuel injection sequences.

Putting EFI and EMS to Work

Any competent back-yard mechanic can convert a carb/points V8 engine to a custom built EFI engine with its own Engine Management System. The most expensive part is the EFI manifold, and if you haven't got a factory version, you can have the holes in the manifold machined to accept EFI injectors (See Converting a Carb Manifold on page 99). To regulate air flow, you use throttle bodies - there are aftermarket types readily available to bolt on to your manifold or you can source an adapter to bolt on some factory GM or Ford throttle bodies.

There are a number of options (See Adapting Throttle Bodies on page 104).

You can even assemble your own computer! The Megasquirt batch injection computer is available from the internet for the price of the components (on average, expect to pay $100 - $150 US Dollars) and you can use an electronic distributor and/or a MSD controller for the fire (the second most expensive item).

FIGURE 8. Megasquirt Computer[3]
Although a fuel only controller, the Megasquirt also has its companions MegasquirtnEDIS - (fuel injection and Ford's Electronic Distributorless Ignition System (EDIS) controller[4])
or MegaSquirt'n Spark, incorporating a simplified ignition control.[5]
You can convert a stack injector (the Hilborn/Enderle mechanical injection used on dragsters) to EFI.

3. http://www.bgsoflex.com/megasquirt.html
4. http://www.jsm-net.demon.co.uk/megasquirtnedis/
5. http://autos.groups.yahoo.com/group/megasquirtnspark/

You can convert a stack injector (the Hillborn/Enderle mechanical injection used on dragsters) to EFI.

FIGURE 9. Stack Injector
Old drag racing engines and their associated Fuel injection components can now be utilised for the street using EFI and a suitably configured ECM. This Jackson unit was picked up at a swap meet and, after a bit of a clean up, will soon be doing duty on a small block powered '55 Chevy.

The sight of eight ram tubes and that fancy plumbing looks the part, but mechanical, constant flow[6] injection isn't going to cut it on a street driven car until you control it with a computer. Swap the constant flow injectors for some modern, pintle type Bosch units, install a few sensors and an off-the-shelf ECM and the entire setup will work as a modern EFI installation.

How about doing the same thing to a flathead? Mike Davidson[7] has been doing it for years. If you want the "nostalgia" look, use a tri-carb setup, only use the carbs to control the amount of air that enters the engine - that's what Throttle Bodies do.

---

6. Mechanical injection is known as Constant Flow. The fuel pump is mechanically driven by the crankshaft and maintains pressure to the Barrel Valve, which distributes fuel at high pressure to each injector. The Barrel Valve also contains the Pill, which senses the engine's demand and adjusts the flow in concert with the opening of the butterflies in the ram tubes. A return line re-directs the unrequired fuel to the fuel tank.
7. http://www.flatattackracing.com/

overview of EMS

FIGURE 10.
Mike Davidson's (Flat Attack Racing)
Hilborn style flathead EFI conversion

Engine Management is simply the total control of our fuel/air ratio and timing while the engine is idling or at wide open throttle. Imagine being able to instantly adjust the jets, float level and accelerator fuel pump on your four barrel carb as you jump on the accelerator. Or imagine re-dialling an advance curve on your distributor every metre of the way up the dragstrip. The simplest systems enable us to do this with ease, and I'm here to tell you your old, carby-equipped small and big block GM, Ford or Mopar V8 can be converted easily and the result is amazing!

But it doesn't stop there!

How about a blown, injected big block that gets reasonable mileage and idles like it belongs on the street? How about adding nitrous to that, and not worry about high speed leanout or pinging or backfiring?

See Controlling Nitrous Oxide on page 37.

FIGURE 11. Big Block Stack Injector
Big block Chevy stack injector, with nitrous injection, that will work beautifully on the street. Idle, cruise and WOT is controlled by the ECM, as is the nitrous. Photo courtesy of Jim's Performance.

Big Block Chevy Stack Injection

There are off the shelf ECMs that cost under $1,000 that can handle 16 injectors AND nitrous AND boost AND cold starts, AND idle speed.

See The Engine Control Module on page 71.

3 Controlling fuel

In this section, we'll look at the components that control the fuel in our MPFI engine. We'll also look at the ways we can control fuel for our engine's idle, cold starts, cruise and Wide Open Throttle (WOT).

The Basics

Power for all gasoline powered engines is a result of fuel burned inside the cylinders. The fuel needs oxygen to burn - we get this from the air.

An engine can be looked at as just an air pump. The amount of power it can make is directly proportional to the amount of air the engine can pump. Building an engine for high performance means designing it to flow more air - simply squirting in more fuel without considering the air flow does nothing.

The optimum air/fuel ratio is 14.7:1, known as the Stoichiometric ratio. For best power, the ratio is about 12.6:1. When building an Engine Management System, we should look at the purpose of the car and set up our fuel accordingly. What this book will focus on is street performance, a balance between all out economy and drag racing. For best results, we need to establish a Stoich mixture (14.7:1) at idle and part throttle (cruise) and at full throttle, the mixture should richen to 12.6:1 for maximum power.

The engine computer stores a map of how much fuel the engine needs at different speeds and throttle positions to achieve the correct A/F ratio. Signals from the Crankshaft Position Sensor and the Throttle Position Sensor tell the ECM what the engine is demanding. The computer looks up those positions on the map and fires the injectors for exactly the right amount of time, measured in milliseconds (thousandths of a second).

Introduction

The brain of any EFI system is the *Engine Control Module* (ECM) which is just a small computer (See *An Overview of Engine Management Systems* on page 19). The ECM receives data from *Sensors* on the engine and transmission, and gives "instructions" to *Actuators* on the engine, transmission, and other parts of the car. For example, the main sensor on any EFI system is the *Oxygen Sensor*

O2 Sensor, See Oxygen Sensor on page 56

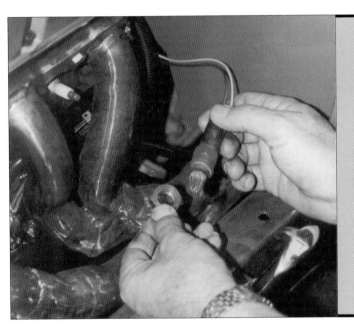

FIGURE 12. Location of the Oxygen Sensor
Oxygen Sensor being fitted to the headers on this hot rod. The O2 Sensor is arguably the heart of the Engine Management System, as this sophisticated device is responsible for reporting the amount of oxygen in the exhaust stream.

If the ECM is the brain, then the O2 sensor is the heart. It tells the ECM how much oxygen is in the exhaust. If there's too much oxygen (a lean condition), the ECM increases the length of time the injectors open, which en-richens the mixture slightly. The O2 Sensor tells the ECM that the oxygen levels are now normal, and the injectors remain at their current state until the oxygen level changes again. This all happens many thousands of times per second, and is just one of the hundreds of controls that the ECM performs.

Open Loop and Closed Loop

Before we go any further, let's quickly look at the two modes of operation that all ECMs will operate in:

Open Loop. The engine is not operating at stoic, that is, 14.7:1 A/F ratio and a pre-determined set of parameters is dictating the fuel and spark. This is usually because:
- The engine is cold, and is in the process of warming up.
- The engine is running at Wide Open Throttle (WOT).

Open Loop may also be activated when there is a fault somewhere in the system, and rather than stop, the engine is running on a pre-determined set of parameters.

This is not the same as when the system is in Open Loop at WOT or when cold, rather it is a "fail safe" mode, sometimes called *Limp Back Mode* or *Limp Home Mode*. In this sense, *Open Loop* means that most of the sensors are being ignored.

Closed Loop. All sensors are working within a specified range, generally at cruise. The engine is warmed up to its normal operating temperature, the Oxygen Sensor is detecting the presence of oxygen in the exhaust stream and relaying the amount of oxygen to the computer, thereby maintaining a near perfect 14.7:1 A/F ratio.

Sensors and Actuators

Sensors are like sender units for gauges – they transfer information about the engine to the ECM. The information provided by the sensors is compared to stored data in the ECM and, if things are not exactly as they are supposed to be, the ECM changes the engine's operation by way of *Actuators*.

There are many different types and designs of sensors. Some are simply switches which complete an electrical circuit while others are complex devices which generate their own voltage under different conditions.

The following components are the more common sensors and actuators that would need to be fitted to a MPFI engine:

Table 1 - Sensors and Actuators

Sensor	Abbreviation	Remarks
Manifold Air Temperature	MAT	Measures air temperature in the inlet manifold
Coolant Temperature Sensor	CTS	Measures temperature of the engine coolant
Oxygen Sensor	O2 Sens, EGO	Measures the amount of oxygen in the exhaust
Heated Exhaust Oxygen Sensor	HEGO	Same as the EGO except that a heating element is added to the sensor
Manifold Absolute Pressure	MAP	Measures the vacuum in the inlet manifold
Barometric Pressure Sensor	BAR, BAP	Measures atmospheric pressure in the engine bay
Throttle Position Sensor	TPS	Measures the angle at which the throttle is open
Vehicle Speed Sensor	VSS	Measures vehicle speed
Engine Speed Sensor	ESS	Measures engine rpm
Crankshaft Position Sensor	CPS	Measures angle of rotation of crankshaft at any given time
Mass Airflow Sensor	MAF	Measures the air as it flows into the engine
Knock Sensor	Knock Sens	Detects detonation in the cylinders
Brake Sensor		Detects application of the brakes
Camshaft Position Sensor	CPS	Checks camshaft position in degrees of rotation
Exhaust Gas Recirculation	EGRVP	Measures position of EGR valve valve position Sensor
Air Charge Temperature Sensor	ACT	Measures the temperature of the air charge entering the motor
Actuators	**Abbreviation**	**Remark**
Fuel Injector	FI	An electrically operated valve, the Fuel Injector squirts atomised fuel onto the back of the intake valve.
Idle Air Control	IAC	An electrically operated stepper motor or solenoid, the IAC provides air for idling when the throttle plates are fully closed.
Electric Fan Drive	EFD	Once the temperature reaches a certain point, the EFD switches the electric fan on.
Wide Open Throttle A/C Shut-off	AC	When you tromp on the GO pedal, the AC Shut Off turns the air conditioning compressor off.
Vehicle Speed Control	VSC	What it says.

controlling fuel

The operation of some of the more important sensors and actuators are discussed in *Sensors and Actuators* on page 53.

Fuel Control

The job of the ECU is to determine the amount of airflow, and pulse the injectors so that the amount of fuel injected provides the desired air/fuel ratio. There are three basic approaches to accomplishing this:

- **Mass Air Flow.** Mass Air Flow systems measure the airflow rate directly. These systems have an air flow meter in the inlet air ducting.

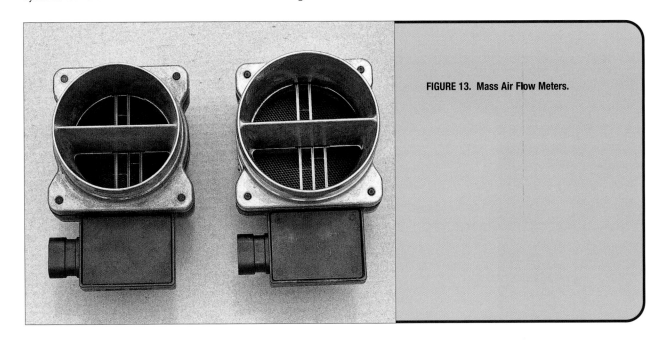

FIGURE 13. Mass Air Flow Meters.

- **Speed Density.** Speed Density systems calculate the amount of air the engine is ingesting by measuring engine speed (that's the *speed* part) and vacuum/boost (that's the *density* part). Throttle position and intake air temperature are also used in the calculation. Speed density ECUs compare these measured criteria against known air flow characteristics of the engine.
- **Alpha-N Fuel Injection.** The *Alpha-N* system of EFI (sometimes referred to as *TPS Mode* or *Speed Alpha*), uses the *Throttle Position Sensor* and engine speed (tacho) to determine engine load. Similar to Speed Density, Alpha-N systems may utilise a MAP sensor for fuel control at cruise speeds or whenever a reliable vacuum signal is present. These systems are useful for engines with very large cams and/or throttle bodies which are exposed to a lot of

air, such as stack injection conversions, twin four barrel conversions that are unstaged or multiple throttle body conversions. The ECU simply responds to the angle of the GO-Pedal[8].

Mass air flow systems:

- Are ultimately more accurate, since deterioration in engine condition and changes in combustion components (cams, blowers, bigger throttle bodies) are automatically accommodated.
- They respond better to changes in atmosphere (elevation, temperature).

Speed density systems adjust for loss of compression (for example leaks past rings or valves) or other departures from new engine specifications only if the EMS has *Self Learning*, a feature normally found on later model, factory cars. However, the ease with which aftermarket ECUs are reprogrammed for changes to a speed density engine compensates for this lack of flexibility.

Speed density systems are at least as satisfactory in practice and have a couple of advantages:

- The flow meter in a mass air flow system imposes a pressure drop which degrades performance somewhat.
- The air inlet ducting must conform to certain requirements for accurate metering.
- There is some built in lag time with Mass Air systems due to the distance between the throttle body and the air meter.
- Throttle response is faster, as the ECU does not have to process data. With speed density, the information is already processed and stored in tables.

It is the simplicity of speed density systems which makes them good candidates for EFI/EMS conversions and the best choice for high performance and race engines.

Engine Control at Idle, Cold Starts, Cruise and WOT

As a general rule, engine management systems control engine operation through four distinct phases:

- Idle
- Cold Start
- Cruise
- Wide Open Throttle (WOT)

---

8. *This method is also employed on 2-stroke engine and rotaries.*

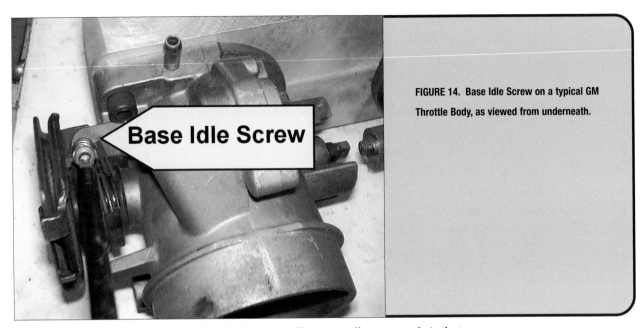

FIGURE 14. Base Idle Screw on a typical GM Throttle Body, as viewed from underneath.

Idle. Engine management at idle is handled by controlling a small amount of air that bypasses the throttle plates. The *Idle Air Control* (IAC) valve or solenoid maintains engine idle speed at closed throttle by controlling the amount of air allowed to bypass the closed throttle plates in the throttle body. See *Idle Air Control Valve* on page 59.

The engine's idle speed is determined by the ECU, and is not mechanically adjustable beyond setting what is referred to as *base idle*, which, in most applications is approximately 500 rpm. Base idle is set by removing the connector from the IAC valve (which eliminates ECM control), and adjusting the throttle plate stop screw (also referred to as the *Base Idle Screw*).

Cold Starts. When an engine is cold, the ECM will cause the injector pulse width to be a little longer than normal. This has the effect of an over-rich condition for ease of starting. At the same time, it will retard the timing, if timing control is incorporated in the system. Early GM TPI systems used a Cold Start Injector, which was basically a ninth injector that came on only when the engine was cold. Most systems today, including after-market ECMs, use the longer pulse width method for cold starts. During the cold start and warming up period, the ECM is operating in *Open Loop* mode, and remains so until the Coolant Temperature Sensor says the engine is warm and the O2 Sensor is warm enough to start detecting oxygen.

Cruise. Once the engine is in *Closed Loop* mode, the ECM determines optimum injector pulse width and ignition timing based on all the inputs coming from all the sensors.

WOT. When the TPS tells the ECM that the driver has stood hard on the accelerator, the injector pulse width changes to a slightly over rich condition and timing is advanced drastically. Some systems go to Open Loop mode temporarily, based on pre-determined fuel and ignition settings. This is because a normal O2 sensor is incapable of determining accurate levels of oxygen in the exhaust due to the temporary over-rich condition. A *Wide Band Oxygen Sensor* corrects this problem, and is used in some high performance applications. These types of O2 sensors can measure O2 levels even at WOT, so that *Closed Loop* operation is maintained.

> *Note:* WB O2 sensors are an advantage only if the software supports the operation during WOT. Most Engine Management Systems go open loop at WOT, or at a pre-determined degree of TPS angle. If the narrow band O2 sensor can be mapped, even at WOT, there is little advantage to using a WB O2 sensor.

Controlling Nitrous Oxide

Nitrous control is far safer and flexible when the ECM is in charge. This section briefly describes the role of the ECM in controlling nitrous and fuel.

FIGURE 15. Nitrous Oxide Injection

How Does Nitrous Oxide Work?

Nitrous Oxide (NO2) is a mixture of two parts of nitrogen and one part oxygen (36% oxygen by weight). When NO2 is injected into the cylinder, it heats up during the compression stroke. As a result, it breaks down and releases its oxygen. More oxygen in the cylinder means the potential is there to burn more fuel, so we inject more fuel at the appropriate time. The nitrogen left behind acts as a dampener which promotes a controlled combustion, counteracting the otherwise more violent flame front that usually accompanies increased cylinder pressure.

Why NO2?

Nitrous oxide injection works well in EFI engines, and most ECMs will accommodate it. It has developed into a popular option for high performance and racing enthusiasts for several other reasons:

- NO2 is a less costly performance modification than a supercharger or a turbocharger.
- Installing NO2 is relatively easy and straightforward.
- Since NO2 is used only when needed, driveability and fuel economy are winners, until you flip the switch.
- Systems can easily be removed or transferred to another vehicle.

Nitrous Management

An EMS that manages a nitrous system will have an extra fuel table which will come into effect when the nitrous activation conditions are met, such as:

- The system is switched on
- The TPS is at a pre-determined angle (minimum throttle position for nitrous activation)
- Engine speed is at a pre-determined level (minimum rpm for nitrous activation)
- The nitrous bottle has pressure

Depending on which EMS you choose, there may be other, or different, parameters. Some systems rely on an external source of fuel, such as a second set of injectors. Usually, however, the extra fuel will come from the existing injectors which will have their pulse width altered to provide the extra fuel while nitrous is activated.

It is critical that the fuel pump, regulator and injectors can supply sufficient fuel when nitrous is being used. Even if your ECM is programmed correctly, if your system cannot meet the physical demands, severe engine damage can occur.

controlling fuel

4 Controlling ignition

It is not uncommon to see an Engine Management System that only controls the fuel. In such a system, ignition advance is set the old fashioned way, and the ECM gets its injector timing from the Tacho signal. Any alterations to ignition advance are via the centrifugal advance system and/or the vacuum advance system. You could even use a points distributor (eeeeeeuuurg!) and it would work fine, provided the input to the ECM was compatible.

In most cases, however, it is advantageous to combine fuel and ignition control in the one engine management system, mainly because the two are so dependant on each other. But first, some revision....

The Basics

To get an engine to run at its best requires the correct air:fuel mixture to be fired by a spark plug at exactly the right moment. It takes a few thousandths of a second from the moment of ignition (the spark) until all the mixture in the cylinder is fully burnt and expanding. Obviously, the spark plugs must fire just before the piston reaches Top Dead Centre (TDC) so that the air:fuel mixture ignites at the right time. *This timing is critical* because the force required to push the piston down and generate power may be weakened by incorrect ignition timing. This time delay is called *Ignition Advance*.

> Note: *We measure ignition timing in degrees of crankshaft rotation, rather than seconds or fractions of seconds of time.*

The perfect time to fire the spark plug depends on engine speed and throttle position. This is a function of the ECU, and operates hand-in-glove with the fuel mixture control. The ECU stores an ignition map of how much *ignition advance* is required in the same way that it stores the fuel map (See *The Basics* on page 31).

The amount of ignition advance required varies from engine to engine. Depending on the engine application, ignition advance should roughly fall between 10 degrees at idle to about 30 degrees at peak revs. Ignition advance increases with engine speed up to about 4,000 rpm and then stays fairly constant (again, this figure is dependant on engine design and application). It increases more rapidly at low throttle openings and slows at full throttle.

Too Much Advance. If the spark plug is fired too soon, the air:fuel mixture starts to burn too soon and tries to force the piston backwards before it reaches TDC - too much advance. This condition kills power and is a potential cause of engine damage.

Not Enough Advance. If the spark plug fires too late, the piston has already moved away from TDC down the bore on the power stroke. The air:fuel mixture is alight and much of the energy released is lost. This condition is known as too much *Ignition Retard*.

Introduction

There are different levels of ignition control depending on the ignition system used. Timing is based on the position of the crankshaft, which was traditionally the function of the distributor. Changes in engine speed and engine load would be managed by centrifugal advance and vacuum advance, while static timing was adjusted by twisting the distributor around. These days, a *Crankshaft Position Sensor* (See *Crankshaft Position Sensor* (CKS) on page 61) and *Camshaft Position Sensor* (See *Camshaft Position Sensor* (CPS) on page 61) are most commonly used as the timing control, giving a far more accurate assessment of piston and valve synchronisation than the distributor. A *Direct Ignition System* (DIS, also called *Distributorless Ignition System* or *Direct Fire Ignition* (DFI)) means that ignition timing is optimised for any condition ñ the ECM triggers the spark. These systems are fairly sophisticated and expensive when converting from a distributor based ignition system, but provide the best spark control and the strongest, hottest spark.

The systems that are described here are just a few of the options available when converting to EFI and EMS. They have been chosen as a guide to enable you to make decisions based on cost, availability (so you can use what is lying around your garage), engine application (street performance, street/strip, drag racing) and level of technology.

Electronic Ignition Management

Essentially, we shall deal with three ignition control systems that are easy to adapt to a wide variety of conversions:

- **Distributor Control.** The simplest way to convert to an EI system that will work with most engine management systems.
- **Crankshaft Trigger Control.** More sophisticated than Distributor Control, but many efficiencies are gained when removing ignition advance control from the distributor.
- *Distributorless Ignition (Coil Packs).* The ultimate in EI, this system will give the best performance, and there are options for the budget minded Hot Rodder who is prepared to do a little fabricating.

Distributor Control

Conversion from a points distributor to electronic is the first step in converting an older engine to EFI. The ECM uses the electronic pulse from the distributor to control ignition and fuel injector timing that a points system cannot provide. The easiest way to achieve this is to use a distributor from a later version of the engine which has electronic ignition (EI). It is usually a straight swap from the old points version to the later EI version.

FIGURE 16. Electronic Distributor
This GM Magnetic Distributor will replace the earlier points distributor and bolt straight in. You'll need the correct trigger box, or ignition module, or you can choose any of a number of aftermarket units, such as the ones supplied by MSD.

Failing this, a custom conversion may be required, using an aftermarket distributor or an electronic conversion kit. MSD, Mr Gasket (Accel), Crane, Mallory etc. have off-the-shelf electronic distributors for just about any engine. Factory systems such as Ford's Duraspark 2, Chrysler's EI (although it would be wise to avoid the Electronic Lean Burn systems, as they were too complex in their day to be as reliable as the Mopar Performance systems, see *Mopar Performance Electronic Ignitions* on page 110[9]) and GMs HEI systems are all good candidates for factory upgrades.

Regardless of which method is used to convert or upgrade, or whether a factory system is employed, the distributor ignition control will always fall into one of the following categories:

- *Magnetic Pick-up.* The distributor contains a Reluctor Wheel and a Magnetic Pick-up, which sends pulses to an ignition module or Trigger Box (often called the Electronic Control Module, but we shall call it the Trigger Box to avoid confusion with the EMS computer). Popular aftermarket Trigger Boxes are manufactured by MSD, Crane, Holley, ProComp (Australia) etc. Factory systems such as the Chrysler orange or chrome boxes or the Ford Duraspark can also be used on engines other than the factory set-up.

---

9. Chrysler's electronic ignition was used on all 1973 and '74 vehicles and on 1975 and newer vehicles that did not use Electronic Lean Burn, or Electronic Spark Control.

- **Hall Effect.** Almost identical to the Magnetic Pick-up, the distributor contains a Hall Effect Sensor and a Hall Amplifier. This generates pulses that fire a Hall Effect trigger box. Hall Effect triggering is less prone to magnetic interference and fires a more precise trigger signal than the magnetic reluctor type. The Hall Effect principle senses the magnetic field strength, not just a change in the magnetic field. It is more precise in sensing magnet position regardless of rotational speed and is less sensitive to variations in gap between the sensor and wheel, making for more accurate timing.

    Note: The trigger boxes could be mounted inside the distributor, which became the norm in later model cars fitted with electronic ignition.

Trigger Boxes

An electronic distributor needs a Trigger Box to amplify the pulse from the distributor's magnetic pick-up or hall effect switch so that it can fire the primary winding of the ignition coil. Some can control ignition advance, provide multiple sparks, have built in rev limiters, etc. Trigger boxes are manufactured to suit all kinds of applications from mild street hacks to dragsters, and are designed to replace the factory trigger boxes and ignition modules, thus bringing multiple sparks and CDI technology without replacing the factory EI distributor. The MSD 6 is probably the most well known, and their performance and reliability are well documented.

FIGURE 17. MSD 6AL Trigger Box

Unlike most factory electronic ignition systems, the MSD employs a Capacitive Discharge Ignition (CDI) that fires multiple sparks under 3,000 rpm. The 6AL has a rev limiter that drops each alternative spark when the engine speed reaches a level determined by one of three supplied plug-in modules. They are still in wide use today, as are similar brands such as Mallory, Holley, Accel and Jacobs, to name a few. For more detailed information on these systems, refer to the book *Auto Electronic* and *Electrical Reference Manual* on page 120. A good ignition system, such as a magnetic pulse or hall effect distributor and a trigger box such as the MSD, will still deliver excellent results when used in conjunction with an aftermarket ECM.

Crankshaft Trigger Control

You can use an aftermarket crank trigger such as the MSD Flying Magnet, Crane Fireball or the Electromotive HPX to determine crankshaft position or adapt a factory crank trigger. The trigger itself comprises a notched wheel and a magnetic or hall effect pick-up. The notched wheel is mounted to the harmonic balancer or, in some systems, the flywheel. The pick-up is positioned so that the notches in the trigger wheel get "sensed" by the pick-up which fires a spark via the ignition module. The air gap between the notched wheel and the pick-up is critical.

Crankshaft position and, therefore, ignition timing, is determined when the trigger box/ignition module senses a control notch on the trigger wheel. This may be in the form of a longer or wider notch, two notches close together or any number of similar methods depending on the manufacturer, however, the result is what is known as the *Reference Angle*. Once the *Reference Angle* is determined, ignition timing is established.

A Crank Trigger can control a DFI system where there is no distributor. See *Distributorless Ignition (Coil Packs)*. Where the distributor is retained, it is only used to distribute the spark, not control it. For this reason, the distributor must be set up to be in phase with the crank trigger. This is known as *Rotor Phasing*. If the distributor has centrifugal advance, you must lock up or weld up the advance mechanism to prevent the rotor phasing from changing with engine speed. Vacuum advance can also be removed and discarded.

Rotor Phasing can be carried out in accordance with the manufacturer's instructions. You must first line up the distributor by eye-balling it, then fine tune it with a timing light.

Coarse Phasing:

1.   Rotate the crankshaft until the #1 piston approaches TDC. Stop when the crankshaft is at the desired timing angle, which is the number of degrees before top dead centre (BTDC) when you want the spark to fire.

2.   Loosen the distributor hold down clamp and rotate the distributor until the rotor is directly aligned with the #1 terminal on the distributor cap.

3.   Tighten the hold down clamp.

Fine Phasing:
4. To observe the rotor, cut or drill a hole in the top of the distributor cap near a terminal.
5. Fit the cap and connect the ignition leads.
6. Loosen the distributor hold down clamp.
7. Start the engine and check the ignition timing.
8. Adjust the ignition timing by sliding the Crank Trigger sensor (magnetic or Hall Effect) in accordance with the manufacturer's instructions. This is usually via an adjustment arm in a slotted bracket.

   Note: Check the sensor to trigger wheel air gap whenever you adjust the timing.
9. Aim a timing light connected to the terminal where the hole was drilled in the distributor cap and observe rotor alignment while the engine is running.
10. Rotate the distributor until the rotor is aligned with the terminal.
11. Tighten the distributor hold down clamp.
12. Install a new distributor cap.

Distributorless Ignition (Coil Packs)

A *Distributorless Ignition System* (DIS) comes in two forms:
- **Direct Fire Ignition (DFI)**. A DFI system has one coil per cylinder. The ECM tells the ignition control module (called the Igniter) to charge the coil and fire it. Spark voltage travels through the spark plug jumping the spark plug gap and grounding itself to the block.
- **Wasted Spark.** A Wasted Spark system uses one igniter and one coil for two cylinders. The ECM tells the igniter to fire the coil on the power stroke and the exhaust stroke simultaneously.

Distributorless ignition systems provide substantial reliability and performance gains over distributors. There is no distributor cap or rotor button to wear out and less gap for the spark to jump (between the cap and rotor). There is also less spark plug wire, therefore less resistance in the high tension circuit. Some direct fire systems place the igniter and coil pack directly on the spark plug, completely eliminating the spark plug wire. DIS systems allow stronger spark and more precise ignition timing control which improves emissions and efficiency, which translates to more power.

Direct Fire Ignition (DFI) is arguably the most efficient and powerful ignition system you can build. It is, however, expensive and sophisticated, and one would have to weigh up the advantages vs cost to justify these systems on a Hot Rod, Custom or EFI conversion project. Most after-market systems rely on a Camshaft Position Sensor to synchronise the location of #1 TDC, which is fine if you have a sequential injection setup (see *Sequential Injection* on page 25). For batch fired systems (see *Batch Injection* on page 25) and/or where a Camshaft Position Sensor is not installed (see *Camshaft Position Sensor* (CPS) on page 61), pickings are slim, but there are options.

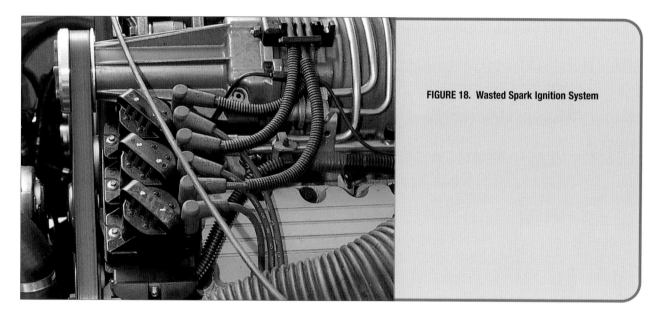

FIGURE 18. Wasted Spark Ignition System

Choosing a Distributorless Ignition System

The fitting of a Distributorless Ignition System is well worth the effort and expense, and there are several options available. Factory system, as a general rule, can only be used with the engine and ECM that came from the donor vehicle. Of course, this is not true if you are prepared to do a little research and fabricating, which we shall see later.

Most of the aftermarket distributorless ignitions provide more options for performance applications than factory systems. Some examples are:

- *MSD CPC System PN 7600.* The Digital CPC System consists of a system controller, ignition control and eight coils. A MSD Flying Magnet Crank Trigger is responsible for triggering the ignition and a special distributor plug delivers a cam sync signal so the controller knows when the number one cylinder is under compression and can start the firing order sequence of the engine. (This plug is also responsible for spinning the oil pump, in most engines.)

- *Holley Annihilator DIS P/N 800-500.* Comprising a High Voltage Module, the Digital Control Module and Ignition Software. Requires a sync signal from a Camshaft Position Sensor.

- *Electromotive HPX direct fire ignition.* No camshaft sensor is required - the HPX relies on a high resolution trigger wheel to synchronise timing. In addition, Electromotive can supply a Total Engine Control system (currently the TEC 3 EMS with DFI) for any engine. This combines their programmable engine management system with their distributorless ignition.
- *Kalmaker and GM Factory Ignition.* Another alternative is to adapt a factory DFI system to your project. One such system that has been used in conjunction with Kalmaker (see Kalmaker on page 81) is the DFI system from the GM 1990 - 1995 LT5 engine as used on cars such as the ZR1 Corvette.

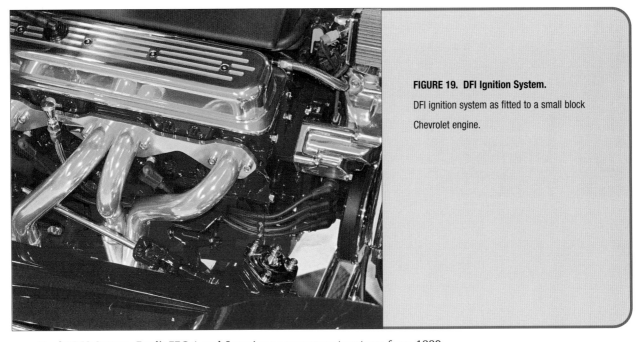

FIGURE 19. DFI Ignition System.
DFI ignition system as fitted to a small block Chevrolet engine.

- *Ford EDIS System.* Ford's EEC 4 and 5 engine management systems from 1990 onwards employed their EDIS ignition control and 259 tooth Crankshaft Position Sensor (they called the sensor a VR - Variable Reluctance, but it means the same thing as Magnetic Pickup). No camshaft sensor was required, as all control was handled by the EDIS ignition module and synchronisation was via a missing tooth on the trigger wheel. Several Programmable Engine Management Systems such as Motec M4 Pro support the EDIS ignition controller, so with a bit of adapting, you can utilise this system on other engines. The Megasquirt and its offshoots (see Megasquirt under Putting EFI and EMS to Work on page 26) utilise the EDIS ignition, too.

Converting Electronic Ignition for use with Delco/Kalmaker ECM

Adapting an aftermarket programmable EMS such as the Haltech, Motec, Electromotive, Hawk, etc to your ignition system is as easy as it can get. Simply select your ignition type from the available options, such as factory magnetic or hall effect distributor, factory ignition module or maybe an aftermarket setup like MSD. Factory ECMs such as the GM Delco, however, are designed for the factory ignition. This may be all well and good if you are using a GM engine and its factory ignition, but what about non-GM engines, or earlier GM points/carb engines?

Kalmaker solves most of the problems because the non-GM components of your EFI conversion can simply be re-mapped to the computer using real values. Some components can be omitted without affecting overall performance, and some components, such as the Knock Sensor, may have to be omitted due to engine noise from forged pistons, roller rockers, timing gears, etc. One area that must be addressed, however, is the ignition.

The Delco EMS is configured for either of the following ignition systems:
- Magnetic Distributor
- Hall Effect Distributor
- Distributorless (Coil Packs)
- Crankshaft Position Sensor (Crank Trigger)

The following procedures describe several ways of adapting GM ignition systems to your project so that you can employ Kalmaker as your EMS no matter which engine you have chosen to power your vehicle.

Magnetic Distributor

If your engine uses a magnetic reluctor distributor, you can connect its pickup coil output to a seven wire HEI ignition module used on small distributor GM TBI and TPI systems as well as the Holden Camira. You can fabricate a piece of aluminium angle to secure the module and act as a heat sink. Bolt the heat sink to the firewall or any convenient place close to the distributor.

Connect the pickup coil output from the distributor to the P and N terminals of the GM ignition module.

> *Note:* You may have to swap the magnetic pickup wires around later.

The ignition module output will then be compatible with the Delco ECM. The pinouts vary between Delco computer models, so consult your factory wiring diagram to ensure the matching four wire interface (EST, Reference, Bypass, and Ref Low) suits the computer model you are using. The following is an example of the 808 or 165 computer:

Table 2 - 808 Computer Pinouts

Module Pin	Module Connector	ECM Connector
E	EST (EST Output)	D4
R	Ref (Crankshaft Position Sensor Reference input)	B5
B	Byp (Bypass Control)	D5
G	Ref Low (Crankshaft Position Sensor Reference ground)	B3

The X and C connections of the module are connected to the + and - connections of the ignition coil. If you are using an aftermarket ignition amplifier, such as the MSD 6 or 7 series Capacitive Discharge systems, connect the X and C terminals to the MSD Red and White wires. The MSD Orange and Black wires are then connected to the ignition coil + and - respectively.

> *Note:* Consult your MSD installation booklet for detailed information on the different configurations or visit the MSD web site to download the instruction booklets.

Hall Effect Distributor

The procedure is similar for a Hall Effect distributor. Connect the three-wire output from the Hall Effect distributor to the GM Hall Effect ignition module (for example, the V8 Holden EFI module). The output to the ECM is the same as for 808 Computer Pinouts.

Kalmaker and Distributorless Ignition Systems

DFI systems are complex and somewhat difficult to adapt to a non-EMS engine, particularly one that will see duty as a high performance street/strip vehicle. Employing a DFI system, however, would provide the ultimate in ignition performance that a distributor system, no matter how powerful, just couldn't. Applying the same theme that has been previously described, we need a DFI system that runs in batch mode, speed density, and requires no camshaft position sensor. The GM 1990 - 1995 LT5 engine utilises the same Delco computer that is found on many other GM vehicles (See *GM Engine Management System ECM* usage on page 115). It also uses a DIS where the ignition advance is controlled by the ignition module, not the ECM. The ECM simply provides the spark timing output, as it would for a HEI distributor. The crank sensor (See *Crankshaft Position Sensor (CKS)* on page 61) is a simple magnetic reluctor device that is easily adapted to other engines or can be fabricated in accordance with the factory GM specifications that are published in the GM technical manuals. The most expensive part of this conversion is the ignition module itself, which costs around $US500 from GM spare parts outlets, however, with a little planning and some Hot Rod ingenuity, nothing is impossible!

> *Note:* There is much more information on this conversion on the Kalmaker web site www.kalmaker.com.au.

sensors and actuators

5 Sensors and actuators

In this section, we'll look at how the sensors are fitted, including custom adaptations for non-factory installations and we'll look closely at a couple of components we haven't examined yet, like fuel pressure regulators and fuel pumps.

Sensors, Actuators

The following topics suggest ways of mounting or connecting the various sensors and actuators. Not all the following sensors are used in all applications. Some systems use only a few sensors, whereas some use others not mentioned here. Check the specifications of the EMS you decide to use and make sure you have all the necessary sensors and actuators fitted.

Mounting of these sensors in non-factory applications is a matter of Hot Rodding ingenuity. When setting up your Hot Rodded EFI system, look at the factory mounting and try to replicate it. Aftermarket components usually come with detailed instructions, adapters and/or bungs that can assist with the installation.

Manifold Air Temperature Sensor.

There are more molecules of oxygen per cubic centimetre of cold air than of warm air, therefore the cooler the intake stream, the leaner the mixture (given the same amount of fuel). The computer has to know about the temperature of the air entering the manifold in order to keep the blend right, and that's the job of the *Manifold Air Temperature* (MAT) or *Air Charge Temperature* sensor (ACT). With a cold engine, the MAT reading should be equal to the outside air temperature.

The MAT sensor can be located in the air filter. It plays a part in determining air/fuel ratios. Some MAT sensors are easily located in the inlet cuff of a typical air filter by drilling a hole and pressing the sensor into the drilled hole, but it must be located somewhere between the air filter and throttle body.

FIGURE 20.
MAT and connector.
Image courtesy of MSD.

Coolant Temperature Sensor (CTS)

The CTS usually mounts in the water passage in the same manner as you would fit a temperature sender unit.

FIGURE 21. CTS.
Photo courtesy of MSD.

FIGURE 22. Location of CTS.
The high performance GM manifold has a provision for the CTS right next to the temperature gauge sender.

Oxygen Sensor

The O2 sensor detects the amount of oxygen present in the exhaust. Given that the best air/fuel ratio is 14.7:1, the O2 sensor can tell the computer whether the engine is running rich, lean or right on target. The ECM adjusts the amount of fuel being injected to keep the air/fuel ratio at 14.7:1. O2 Sensors do not start working until they are hot, so while the engine is warming up, the system is in "Open Loop" mode. Until the O2 sensor is heated up and working, the engine assumes that the mixture is okay. O2 Sensors can be heated quickly by the battery (two or three wire sensors) or by the exhaust temperature alone (single wire sensors).

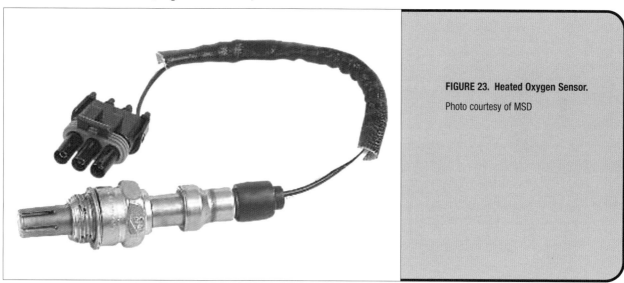

FIGURE 23. Heated Oxygen Sensor.
Photo courtesy of MSD

The O2 sensor generates its own voltage (typically between 0mV and 1000mV) depending on the amount of oxygen present in the exhaust. At 14.7:1 A/F ratio, a typical O2 sensor will generate 450 millivolts (mV, or thousandths of a volt). A lean mixture (more oxygen in the exhaust) will produce a voltage below 450mV, and a rich mixture will produce a voltage above 450mV (less oxygen in the exhaust). When the EMS is running in closed loop, the O2 sensor will vary constantly above and below the 450mV mark. The ECM can then maintain an average A/F ratio of 14.7:1.

O2 Sensor Mounting. The O2 Sensor must be mounted close enough to the exhaust ports to ensure that it heats up as quickly as possible. For headers, a bung can be welded into the collector and the sensor screwed into place. For manifolds, the same technique can be used at the closest point to the exhaust manifold flange as possible.

Take a note of where the O2 sensor is fitted in a factory setup that is similar to your own project as a guide.

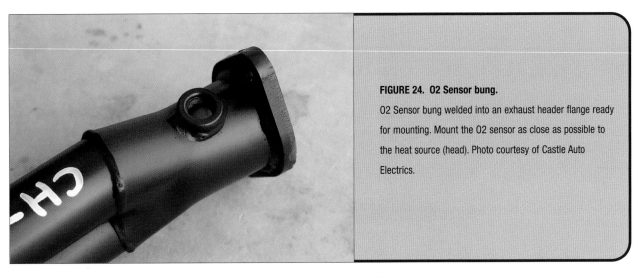

FIGURE 24. O2 Sensor bung.
O2 Sensor bung welded into an exhaust header flange ready for mounting. Mount the O2 sensor as close as possible to the heat source (head). Photo courtesy of Castle Auto Electrics.

Heated Exhaust Oxygen Sensor (HEGO). Heated O2 sensors are two or three wire units that have their own heat source via a +12V supply. Use one of these in preference to the single wire unit. The same mounting procedure applies.

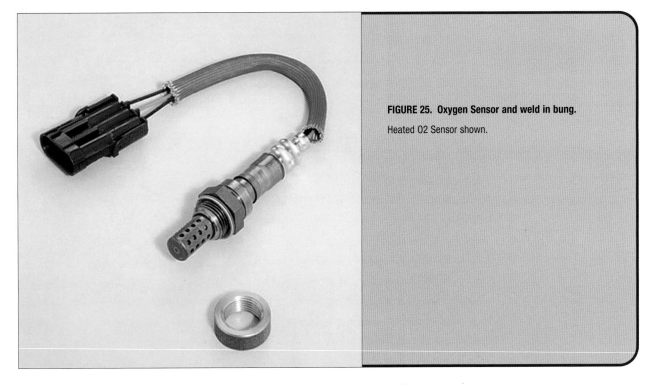

FIGURE 25. Oxygen Sensor and weld in bung.
Heated O2 Sensor shown.

Warning: Leaded fuel, anti-freeze, vapours from silicone sealants and long exposure to a rich mixture can destroy an O2 Sensor.

Manifold Absolute Pressure Sensor

The Manifold Absolute Pressure Sensor (MAP) measures vacuum inside the intake manifold. The ECM receives information on engine load via this sensor. High pressure (low vacuum) indicates a heavy load and high power output. The ECM issues commands to enrichen the fuel mixture and decrease the spark advance. The opposite occurs for high vacuum conditions, where there is little load. A leaner mixture is sufficient and more spark advance is required.

FIGURE 26. MAP Sensor.
Note the barbed vacuum connection that connects a vacuum line to the manifold. Photo courtesy of MSD.

The MAP is used in *Speed Density* systems to help calculate engine load. They are often located inside the ECM, and you simply hook a manifold vacuum hose to the barbed end. They can be mounted just about anywhere.

Warning: Some vacuum lines are made of synthetic rubber that conducts electricity. DO NOT lay manifold vacuum lines across the printed circuit board of the ECM!

MAP sensors are available in 1 Bar, 2 Bar or 3 Bar sizes. The ECM reads manifold pressure as an absolute value, that is, as the pressure inside the manifold, not as a vacuum with respect to outside atmosphere (Barometric pressure). For normally aspirated engines, the 1 bar MAP is fine. For turbo and supercharged engines, the 2 and 3 Bar MAPs are required, as they have to measure higher absolute pressure than the vacuum inside a normally aspirated manifold.

Barometric Pressure Sensor (BAR, BARO, BAP).

The BARO sensor detects barometric pressure of the atmosphere. The ECM uses BARO to help determine engine load, and, therefore, air/fuel ratio. The BARO sensor responds to changes in altitude, and the ECM can make adjustments accordingly. MAP and BARO sensors are often combined. Like MAP sensors, too, they can be mounted inside the ECM. BARO sensors are not used much any more because the ECM can use the MAP to read BARO at engine start up.

Throttle Position Sensor.
The *Throttle Position Sensor* (TPS) allows the ECM to sense the position of the throttle and to use that information to control fuel delivery. The TPS is simply a variable resistor whose output is dependent upon the position of its rotating wiper arm which is connected to the throttle shaft on the throttle body. At idle, the throttle plate in the throttle body is closed, and the air that's breathed by the engine is supplied by the *Idle Air Control* (IAC) Valve. When the accelerator is depressed, the throttle spindle rotates and the plate "opens" to admit more air. The TPS tells the computer exactly where the throttle plate is at any given time.

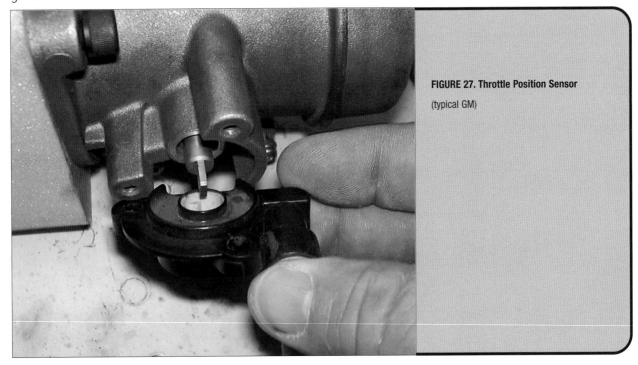

FIGURE 27. Throttle Position Sensor (typical GM)

The TPS is usually a three-wire sensor mounted to the shaft of the throttle body. Throttle position is used to determine A/F ratio, spark timing, idle speed, torque converter lockup (TCC), A/C clutch, and is the primary controller in an Alpha-N system.

Using factory throttle bodies means that you already have the TPS and Idle Air Control valve fitted. If you are using modified carb bases, stack injection or a custom built throttle body, you need to consider ways of mounting the TPS. The solution can be to mount it on the accelerator shaft. This has the added advantage of hiding the TPS and its associated wiring.

FIGURE 28. TPS in place.

Idle Air Control Valve

The Idle Air Control (IAC) valve can be a small stepper motor which operates an adjustable tapered valve.

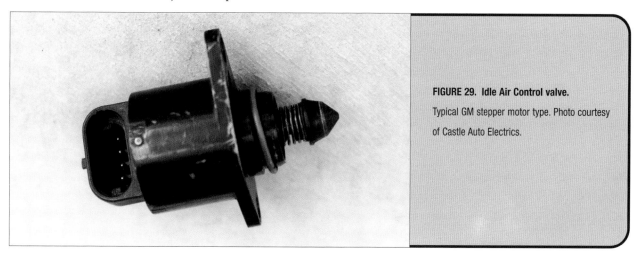

FIGURE 29. Idle Air Control valve.
Typical GM stepper motor type. Photo courtesy of Castle Auto Electrics.

Some IAC valves are simple solenoids, and are either ON or OFF. The valve or the solenoid maintains engine idle speed at closed throttle by controlling the amount of air allowed to bypass the closed throttle plates in the throttle body. The engine's idle speed is determined by the ECU, and is not mechanically adjustable beyond setting what is referred to as base idle, which, in most applications is approximately 500 rpm.

Where no IAC mounting exists on the throttle body (such as stack injection or when using carb bases as throttle bodies) a remote IAC valve can be used. All you need is a vacuum line. See *Remote IAC Valves* on page 112 for information on where to obtain a remote IAC valve.

FIGURE 30. Remote IAC Valve
A common aftermarket IAC Valve is mounted on an aluminium block. Simply connect one outlet to a vacuum source (below the butterfly on the throttle body) and the other to a fresh air supply.

Engine Speed Sensor (ESS).

The ESS monitors engine speed and, in some applications, crank position. It is used to calculate injector pulse width. There are various factory and non-factory types that can be mounted on the flywheel housing, harmonic balancer, etc. The ESS is rarely used in non-factory applications, as engine speed can be easily determined by the Tacho signal.

Vehicle Speed Sensor (VSS)

The Vehicle Speed Sensor (VSS) sends a signal to the ECM which the ECM converts to a speed reference. This sensor mainly controls the operation of the automatic transmission's Torque Converter Clutch (TCC) system as found on GM T700 transmissions. It also provides an input for anti-lock brake systems, electronic cruise control and the electronic speedometer.

Note: The VSS also determines deceleration enrichment in some Engine Management Systems by controlling the opening and closing of the IAC. Where this is the case, the VSS has a profound effect on driveability, and should be retained in your custom EMS to prevent stalling and to smooth out the transitions from deceleration to WOT.

On GM cars, the VSS is mounted directly in the transmission extension housing, where the speedometer drive gear sleeve would have been. If your car uses a speedometer cable, the VSS can be a small unit mounted to the back of the instrument panel behind the speedometer and driven by the speedometer cable. In most Street Rod or similar installations this function is not required, unless you would be interested in having the TCC function totally as originally designed, or adapting a late GM electronic cruise control to the installation.

Crankshaft Position Sensor (CKS). A CKS is mounted close to a toothed wheel (or a notched wheel), usually on the harmonic balancer, but can also be mounted on the flywheel/flex plate. As each tooth moves past the sensor, a pulse is induced in it, which is sent to the ECM. Each tooth produces a pulse. The number of pulses per second is the *Frequency* of the signal. The frequency translates to engine speed, which is used to time the ignition.

Camshaft Position Sensor (CPS). With assistance from the *Crankshaft Position Sensor*, the CPS tells the ECM where #1 TDC is. Because the engine's firing order is programmed into the ECM, it can use #1 TDC as a reference point to fire coil packs in a *Direct Ignition System* and/or to synchronise injector firing for SEFI systems. The CPS is similar to the CKS in construction, although some use a toothed wheel with a longer tooth or a double tooth to mark #1 TDC. In some applications (eg, on Toyota quad cam V8s) this sensor is called the Variable Valve Position sensor

Mass Air Flow Sensor. The *Mass Air Flow Sensor* (MAF) measures the volume of the air entering the engine. This information is used by the computer to supply the correct amount of fuel. There are several types of MAF sensors, but the most common is the *hot wire* type that determines air flow by measuring the current required to maintain a heated wire at a constant temperature as intake air passes over the wire. The MAF sensor is the chief compensating feature in Chevrolet's TPI. This unit must be mounted between the air cleaner assembly and the TPI throttle body. In fact, regardless of make, a sensor between the air cleaner and the throttle body is a sure sign that the ECM is Mass Air and not Speed Density.

sensors and actuators

FIGURE 31. Mass Air Flow Sensor.
GM LT1 Shown.

Knock Sensor (Knock Sens). The knock sensor, located on the engine block, detects vibrations that are the acoustic signature of detonation. The Knock Sensor informs the ECM that detonation is occurring, and to retard the timing in an attempt to eliminate the detonation. Up to 20 degrees of timing can be retarded to compensate for bad fuel, high engine temperature, or any other combination of factors that produce detonation.

Idle Air Control (IAC). The IAC valve is an electrically operated valve or stepper motor which regulates a volume of air bypassing the closed throttle body butterflies. The IAC establishes a stable idle when the engine is cold and maintains idle speed when it has warmed up.

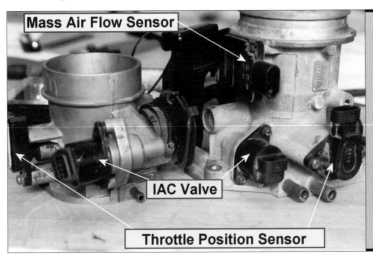

FIGURE 32. IAC Valves.

This is how an IAC valve is mounted in GM Throttle Bodies. The throttle body on the right is a Mass Air Flow type. The one on the left is a Speed/Density type from VN Commodore. When custom building a manifold and throttle body set-up, you may want to consider a remote IAC.

The stepper motor type of IAC valve bypasses an amount of air determined by the number of 'steps' the motor takes – the more steps, the larger the air flow opening and the more air is bypassed. The IAC can also be an On/Off type, which bypasses a set amount of air when it is activated.

Fuel Pump Relay

The fuel pump relay is controlled by the ECM and acts as a remote switch to route power to the electric fuel pump. A relay is necessary because most EFI fuel pumps can draw up to approximately 10 amps of current.

Fuel Pressure Regulator

There are many different EFI fuel pressure regulators, but their purpose is the same - to maintain fuel pressure at a certain value above the intake manifold pressure. The inner mechanism usually consists of a sealed diaphragm, a spring, bypass valve and a manifold pressure reference port (a vacuum line). The bypass valve is connected to the diaphragm and the spring pushes against it from the manifold pressure side. The spring pressure determines the base fuel pressure. If there is vacuum at the port (eg, at idle), the spring pressure on the diaphragm is less, therefore the fuel pressure under vacuum conditions is lowered. If there is pressure on the port, such as under boost, spring pressure increases, as does fuel pressure. Most regulators have a static pressure of between 38 and 44 psi.

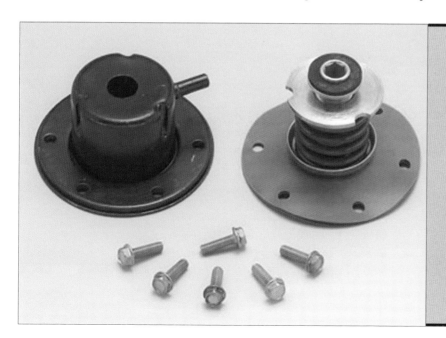

FIGURE 33. Fuel Pressure Regulator.
Image courtesy of Accel.

The fuel pump is always producing an excess of fuel volume. Once the base pressure is met, the regulator controls the pressure in the fuel rail by bypassing unused fuel back to the fuel tank. At idle, about 95% of the fuel delivered to the fuel rail is returned to the tank. At full power, somewhere between 5% and 50% of the fuel is returned to the tank. You should route the fuel from the pump to one end of the fuel rail to feed the injectors. The regulator is mounted on the opposite (low pressure) end of the rail. Remote mounted regulators can be mounted on the firewall and plumbed with high pressure fuel line. Mounting the regulator on the low-pressure side allows any hot fuel in the rail to be purged back to the tank after a hot start to reduce vapour lock and fuel boiling.

FIGURE 34. Adjustable Fuel Pressure Regulator.
This is a Holley regulator, a new product. It can be mounted just about anywhere. Image courtesy of Holley.

For supercharger or turbocharger applications, you'll need a Rising Rate regulator (sometimes called a Fuel Management Unit (FMU)). Rising Rate regulators increase fuel pressure at a multiplication factor of boost. Instead of computer tuning and thrashing injection duty cycles, these systems increase fuel pressure to add fuel. They are installed downstream from the normal regulator and only start to add pressure under boost. When the engine is off boost, normal tuning and drivability is maintained.

Warning: Setting the fuel pressure regulator to extreme pressures (over 60psi) can cause your engine to run extremely rich without you knowing it. It can cause serious fires in catalytic converters or fuel up your exhaust system, destroying your O2 sensor and creating a fire hazard. Be careful!

Fuel Injectors

There are so many different types of fuel injectors around that it would be impossible to document them in the space available. Described here are the basics, which should be used as a guide for injector choice.

Pulse Width and Duty Cycle. The length of time an injector is open and squirting fuel is called the *pulse width*, and is measured in milliseconds (ms). That's thousandths of a second. As engine speed increases, an injector can only be held open for so long before it needs to be held open again for the next engine revolution. This is what is known as the injector's *duty cycle* and is measured as a percentage of open to closed.

Injector Size. Injectors are measured in accordance with their Flow Rate in lbs/hr (pounds per hour) or cc/min (cubic centimetres per minute). A fuel injector's flow rate is measured at its maximum duty cycle (100%), but in real life they should never be operated at more than 80%. The following formula is used to "guestimate" injector size for a particular application:

BSFC = Brake Specific Fuel Consumption, generally accepted to be around 0.5 for calculating injector sizes.

$$\text{Injector Flow Rate (lb/hr)} = \frac{\text{Engine HP} \times \text{BSFC}}{\text{Number of Injectors} \times 0.8}$$

For example, to calculate the individual injector size for a 650 HP V8 using 8 injectors and assuming a BSFC of 0.5:

$$\text{Injector Flow Rate (lb/hr)} = \frac{650 \times 0.5}{8 \times 0.8} = 50.78 \text{ lb/hr}$$

To convert lb/hr to cc/min, multiply by 10.5.
For more details on Injector Flow Rates, see *Flow Rates/ Pressure* on page 67.

Injector Types

For simplicity, these articles describe injectors as being one of two different types according to their internal electrical resistance:
- **Saturated.** High resistance. Saturated injectors are used in almost all standard

production, engines, mainly because of the cost and simplicity. Saturated Injectors are more suited to Batch Injection.

- **Peak and Hold.** Low resistance. Peak and hold injectors respond faster than Saturated injectors. They require more electrical power to open (4-6 Amps vs. 0.75 -1 Amp). Peak and Hold injectors squirt fuel more accurately, and sustain the accurate squirt for longer, than the saturated type. This is important in high-pressure systems. Using Peak and Hold injectors costs more because they require one computer injector driver per injector in most applications. This is a design requirement in Sequential Electronic Fuel Injection (SEFI) systems where each injector is fired at a very precise time in the intake cycle.

Most mass produced cars use Saturated 12 to 16 Ohm Injectors. Peak and Hold injectors are generally reserved for high performance applications or in Turbo/Blower applications.

FIGURE 35. Fuel Injectors

Fuel Injectors come in many sizes and types, but are generally divided into two groups – Low impedance (Peak and Hold Type) and High Impedance (Saturation). Injector drivers are rated differently for the two types, so care must be taken to fit only those injectors that are compatible with your ECM. Fitting the wrong injectors may damage the ECM or, at best, will perform poorly.

Injector Choice

Choosing an injector means selecting the correct flow rate and the type supported by your ECM. See *Injector Size* on page 65 for details on the different types and how to determine the correct flow rate. Also, Table 9, *Fuel Injectors*, on page 118 lists fuel injector data for a number of different injectors and Table 10, *Injector Flow Rates*, on page 122 lists injector flow and horsepower to assist with injector selection. The following are things to consider for your EFI project.

Fuel Transfer and Sealing

Transferring fuel from the fuel rail to the injectors is via O-ring seals. Sealing injectors to the intake manifold is similar, typically a 14mm round section O-ring sitting in an isolated groove in the manifold or the injector bung. Securing the injectors can be achieved by simply holding the fuel rail against a bracket welded to the manifold.

FIGURE 36.

Securing the Injectors.

On this Stack Injection conversion, a bracket has been welded to the fuel rail and bolted to the throttle bodies. Simple, but practical, it should clean up nicely. The injectors haven't been fitted yet. Photo courtesy of Inner Active Manifolds, Blacktown.

Factory installations usually utilise a clip at the base of each injector, but no matter which method you choose, they must be held securely in place in case of a back-fire. Under normal conditions, the O-rings will hold them in the manifold just fine, but sudden surges in fuel pressure, too much overlap from your lumpy cam or backfire may dislodge the injectors unless they are held in place. Blower or turbo systems will also cause a fair amount of back pressure, so be aware of this when setting up a method of securing your fuel rails.

> *Note:* You should always replace the O-rings on used injectors before securing them. Smear some engine oil on the O-rings first.

Flow Rates/Pressure

Most factory injectors are quite small because of the lower power production engines. Fuel metering is also more precise with small injectors and they are also better for smooth idle and emissions. Very few production engines use an injector flowing more than 500cc/minute or 50lbs./hr. For performance applications, engines often require much

FIGURE 37.
Securing Fuel Rails.
The fuel rails on this 351 Ford Cleveland are secured by a couple of posts that are actually two of the manifold retaining bolts. Very clever, very easy and very neat. Photo courtesy of Inner Active Manifolds, Blacktown.

larger injectors to satisfy the increase in fuel flow. Often larger factory injectors can be fitted from a different engine. Sometimes aftermarket ones must be used. Here are a few things to consider when deciding on injectors:

- Choose an injector large enough to feed your engine at maximum power. See Table 10, Injector Flow Rates, on page 129.
- Use factory injectors where possible ñ they are much cheaper and more readily available. Most factory EFI systems maintain a fuel pressure of between 36 and 43.5 psi over the intake manifold pressure, so use this as a guide.
- Table 9, Fuel Injectors, on page 125.
- Fuel pressure can be raised to increase the rate of fuel flow. To double fuel flow, you must increase fuel pressure by a factor of four. Do not exceed 60 psi in most cases. Raising the pressure too high will hurt the fuel pump and can lead to leaks or failures in the plumbing and injectors themselves. Use the proper flow rate for the intended application.
- Injectors for performance engines should be flow and leak tested first. If they are not in peak condition, the engine will never run well.

sensors and actuators

the engine control module

The engine control module

In *An Overview of Engine Management Systems* on page 19, we said that batch systems are far less complicated from a software and hardware standpoint. It stands to reason, then, that the engine management systems will be less expensive to purchase and easier to program. SEFI systems, on the other hand, require timing input to the ECM either derived from a *Camshaft Position Sensor* and/or from multiple input signals from the *Crankshaft Position Sensor*. Because each injector is timed, SEFI computers require a separate driver for each injector, a separate wire for each injector, stored information such as cam timing, injector response times, firing order, etc, not to mention the software to process the data. Programming the ECM will require higher levels of computing skills so that all these aspects, which inter-relate to each other, can be brought together.

Which System?

Batch or Sequential? Mass Air or Speed Density? Factory or after-market ECM? Because every project is different, it will be difficult to decide which EFI system will work best for a particular engine - it depends on the application of the car. The focus might be on performance rather than economy, or daily use vs weekend use. As with any system, when performance is a priority, engine speeds are going to be higher and fuel flow rates are going to be higher. This is a most important aspect to consider when putting all the EFI components together.

As engine speed increases, the amount of time available to inject the fuel decreases.

The batch-fired system is injecting fuel for the entire period of crankshaft rotation. However, the only time allowed for the SEFI system to squirt fuel is when the intake valve is opening. To compensate, SEFI systems must have larger injectors to achieve the engine's rated horsepower. Also, air velocity is low in the port and runner until well after the valve is open. If the injection pulse starts early and finishes late to compensate, then we have lost the fine-tuning advantage that SEFI provides in a less performance-oriented engine. On a high revving engine, a scant few milliseconds are all that's available to inject fuel before the next cycle begins.

With batch injection, fuel is squirted at the back of a closed intake valve at least once per cycle. Most *Batch Injection* systems are actually *Bank Injection* (sometimes called *Group Injection*), where the cylinders are divided into two or four Banks, depending on the firing order (that is, during the first cycle, one bank is fired. During the next cycle, the next bank is fired). As engine speed *increases*, the time that the fuel sits in the port *decreases*, so this action has little bearing on wide open throttle performance. In reality, there is little difference in fuel economy between the two systems, although emissions at part throttle are better with SEFI, but this is what it was developed for in the first place. To be perfectly candid, performance applications of EFI disregard emissions in the same manner that we disregarded them with carbs. When you bolt a tunnel ram to that V8 and run a half inch cam, you aren't doing it to get fuel economy! However, converting to EFI, even for the wildest engine, means lower emissions across the rev range of the engine, so we can still get a warm, fuzzy feeling between bursts of 12 second passes. Remember also that if your car is not equipped with a catalytic converter, it will be difficult (but not impossible) to reach today's emissions standards anyway.

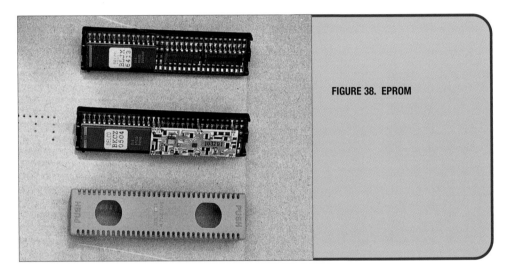

FIGURE 38. EPROM

Programming your EMS for Engine Modifications

Firstly, let's get one thing perfectly clear – for most naturally aspirated engines operating on pump fuel, the only way to achieve real power gain is by increasing airflow through the engine. If you are going to install a factory EFI engine into your Hot Rod or Street Machine, then "fiddle" with the electronics to get it to go hard, think again! *Chips cannot do this.* A Chip is simply an electronic component known as an Integrated Circuit (IC). They are called "Chips" because the material they are made from is a chip of silicon. What are normally referred to as chips in ECMs are actually storage devices, such as the EPROM (Erasable Programmable Read Only Memory), EEPROM (Electrically Erasable Programmable Read Only Memory) and SRAM (Static Read Only Memory). These chips store information about the engine, its sensors and actuators and the computer program that manages the data. In factory systems, access to the data and the computer program stored on the chips is locked, and protected by copyright and/or license. Programmable engine management systems, however, are empty shells that you fill with your own data, including the program that you are licensed to use.

Aftermarket chips. You can buy chips that you plug into a stock, factory ECM to "hot up" your stock engine. These chips have been re-programmed to richen the mixture slightly at full throttle and/or advance ignition timing, but at the expense of lowering the safety margins for detonation and emissions set by the factory. You must also run a higher octane fuel than that recommended by the factory - premium unleaded at least.

There are also aftermarket chips for modified engines, often called *Custom Chips*. These are fine if your engine has the same modifications as the engine used on the chip company's dyno – that's what the chip was developed for. Some performance parts warehouses sell a complete package – cam, pistons, heads, headers, throttle bodies and chip. However, if your cam, heads, exhaust, injectors or throttle body are different, the chip will not work correctly in your engine. The only way to get EFI to work right on a modified engine is to take your engine to someone with the right facilities and get a custom chip burnt for it or use an engine management system that is fully programmable. This is a process that starts by measuring all your engine parameters (preferably on a chassis dyno, but not essential) transferring the data to your ECM, then testing on the road. Repeat until it's perfect. You can do this at home if you have the patience, a long stretch of deserted highway and you are computer savvy. Being a propeller head is an advantage.

So where does that leave the majority of us?

Your choices are to use an aftermarket *Programmable Engine Management System* (some examples are Electromotive, Haltech, Motec, EM4), or a factory ECM that comes with the stock factory engine.

Introducing Kalmaker

Kalmaker, on the other hand, is licensed software you can purchase that works in concert with the unadulterated GM program and your custom data in one of a number of compatible GM Delco ECMs. No matter whether your engine is a souped-up 1999 LT1 or a 390 Caddy, they can all be electronically managed in this way.

Kalmaker turns the GM/Delco ECM/PCM into a Programmable Engine Management System. More on Kalmaker later.

Choosing an Engine Management System

Manufacturers of aftermarket programmable engine management systems have vastly simplified the programming of their computers so that even a trained monkey can install, wire, and program the most sophisticated SEFI engines. The software that is supplied is usually written to accommodate multiple configurations, and can be set up as batch, bank, sequential or Alpha-N (or *TPS Mode*, see page 35). You can also select speed density or mass air. You can set it up for a mild, economical grocery-getter or as an all out race car. The software that drives these systems is designed to be as intuitive and user-friendly as possible, and it accomplishes this by only providing tuning for a set number of essential engine parameters. Compared to factory computers like the GM TPI or Ford EEC systems which control something like three hundred engine parameters, the programmable EMS may only control twenty.

Features such as closed loop idle, closed loop cruise, accelerating or decelerating using a Vehicle Speed Sensor (VSS) may be ignored in aftermarket programmable EMS. They may not include air temperature, manifold air temperature, water temperature or barometric pressure in their calculations, again, to provide for ease of installation and setup. The number of load points, too, will be far less than those for factory ECUs. What this means is that there are refinements that may or may not be available, such as:

- Tuning out backfiring/popping condition when slowing down/backing off.
- Timing adjustments that prevent pinging on hot days yet still make maximum power on cold days.
- A steady reliable idle regardless of the conditions.
- Starting first time, every time without touching the throttle, whether the engine is hot or cold.
- Low speed, first gear driving without surge or having to ride the clutch.

To put this in perspective, the best programmable EMS is capable of addressing up to 20 maps. The GM Delco ECM has about 300 maps!

There are a myriad of non-factory *Programmable Engine Management Systems* to choose from. All of the ones listed at the *Appendix* on page 113 provide some means of programming from a hand held device or a laptop. A degree of computer literacy is usually required. While the programming has been considerably simplified, it means that the overall control of the engine is going to be fairly basic compared to factory systems. If simplicity and ease of programming are major issues for you, an aftermarket programmable EMS may be the answer.

Factory computers build many sophisticated controls into their software, some of which are copied in aftermarket systems. The following are some controls and sequences that you should consider when choosing an engine management system for your project:

- Limp Home Mode. What happens when the ECM fails, for instance?
- Full Real Time Editing and total reprogramming capability.
- Full Real Time Diagnostics with data logging of engine inputs and outputs.
- Use Speed Density or Mass Air.
- Use TPS mode in configurations such as stack injection conversions or race engines

with very big cam angles. Program the ECM to ignore the MAP sensor when vacuum is unreliable.
- Closed Loop Fuel Control and Idle Control.
- 1, 2 or 3 bar MAP sensor can be used for forced induction systems.
- Twin Tables for different fuels.
- Self Tuning VE Fuel Update.
- Single Fire injector pulse for idle quality for large injectors.
- Knock Sensor Control which can ignore noises such as roller rockers and forged pistons.
- MAT (manifold air temp) Spark Correction.
- Water Injection either direct off/on or duty cycled to suit Aquamist etc.
- Closed Loop Electronic Wastegate Control.
- Full mapping to 9600 rpm.
- Adjustable Rev limiter.
- Boost cut.

The only factory EMS that is dealt with in this document is the GM Delco that is compatible with Kalmaker. You can purchase a Kalmaker system or have a licensed Kalmaker operator modify a compatible Delco computer for you. Using Kalmaker, you can have the advantage of the intensive and costly research and development undertaken by GM/Delco as well as the flexibility to adapt the system to just about any engine with almost limitless modifications.

Kalmaker

OEM computers are built for specific engines and programmed for specific applications. You can't just grab the first computer you find in a wrecking yard and wire it in to your 351 Ford powered '36 Coupe or your 390 Cadillac powered Hi Boy roadster. It would be cheaper, though, wouldn't it? More importantly, factory computers control many more engine parameters than aftermarket programmable engine management systems and the software is far more sophisticated.

Why?

Well, manufacturers are required to meet stringent anti-pollution standards, which can only be obtained by operating their engines to very tight specifications. In addition, it is far cheaper to design a sophisticated engine management system to squeeze maximum efficiency out of an otherwise ordinary engine. Which makes modifying an engine a bit more difficult, as the EMS will require modifying, too. There are, however, some exceptions.

the engine control module

Mass Air systems such as the '86 - '89 GM TPI will give you a lot of leeway for engine modifications. You can build your 350 with a lumpy cam, alloy heads, balance, port job even plonk a blower on, and the factory ECM will adjust accordingly without any intervention by you. Provided that the injectors will flow enough, the ECM will only see that the engine is ingesting a larger amount of air than normal, increasing injector pulse width accordingly. Later models (post 1990) are speed density and combinations of speed density and mass air. More importantly, the ECM is programmed with more specific data, which relies on engine specifications remaining constant.

As previously stated, most factory computers are designed for a specific mass produced engine, with known, fixed parameters such as cam timing, combustion chamber volume, ignition advance, etc. They also rely on sensors that respond in a consistent manner, providing the ECM with the measurements consistent with the software. The only sure way to ensure your engine will work using a factory EMS is to keep the engine and transmission exactly as it came from the donor vehicle. A factory EMS will not, as a general rule, tolerate engine modifications. It may still work, but not as efficiently.

For us Hot Rodders, factory EMS salvation comes in the form of an Australian software company whose founder is a performance car enthusiast. Ken Young designed and wrote Kalmaker[10], a computer program that "talks" to the common GM AC Delco engine management system.

In Australia, the 1987 to 1996 Delco computers are all candidates for Kalmaker. Delco ECMs are used on Commodores, Camiras and Astras. They are also used on the 1986 – 1989 USA version for Chevy TPI[11] and similar computers for 90 - 96 TPI.

For part numbers, identification details and applications, see *GM Engine Management System ECM* usage on page 115. The good news is that these cheap, plentiful computers can be customised for any engine and any application – from street use to drag racing – simply by hooking a laptop computer to it and changing a few values. You save the values just like you'd save any computer file and then drive away.

In this section, we shall use the GM Delco computer and the Kalmaker software as an example of how a custom Engine Management System is programmed. The process is the same for any EMS, but the Delco/Kalmaker combination covers every possible engine control and parameter, so no matter which system you use, following the Kalmaker tuning process will ensure that all aspects of engine management are covered.

---

10. *Kalmaker was originally derived from DynoCal, an older professional version developed by Ken Young*
11. *See my book, the Small Block Chevy TPI System.*

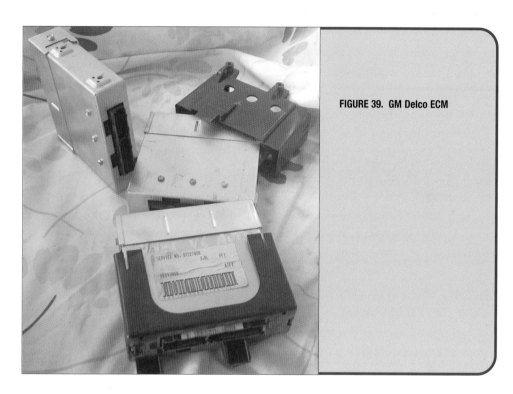

FIGURE 39. GM Delco ECM

The Software

The ECM is simply the platform that runs the engine management software. The better the software, the better the engine management. Car makers have invested a lot of research and development and spent millions perfecting engine management software. Kalmaker turns the Delco software into a format that we can read on a laptop computer, thereby giving the user the power to re-map every section of the factory engine management system. Aftermarket engine management systems such as Haltech, Motec, Electromotive, Hawk, to name a few, also give you the power to control the engine's operation, but there are far fewer control points to "map", thus simplifying the tuning process. Using Kalmaker in your EFI conversion, while giving you benefits not found on aftermarket programmable engine management systems, means more effort is required to "fine tune" the software.

Therefore, you should weigh up the advantages:
- Low initial cost and low maintenance costs. The computer, sensors and other such components are easily sourced from wrecking yards, swap meets and GM parts outlets.
- The system comes with many factory calibrations so that you don't have to start from scratch.

- Your vehicle can be serviced or analysed by a GM Dealer in the event of a breakdown. Factory manuals are still relevant.
- The Service Engine Soon light will flash trouble codes just like the factory intended, without having a laptop connected.
- Self Learning adjusts the engine's optimum parameters as the engine ages.
- Fuel economy is improved, especially with Lean Cruise mode selected.
- Deluxe Idle Control - PN/Drive step up, AC step up and Fan step up are all included.
- Decel fuel (fuel is almost shut off when decelerating) to save catalytic converter and improve emissions.
- Pass the most stringent emission tests, if required.

The end result is improved driveability, the biggest benefit over all other aftermarket systems.

Using the Delco ECM and Kalmaker

So what does a rodder have to do to get a Delco ECM re-programmed? There are two ways:
- Purchase the software (Street Pro 3) to suit your application and tune it yourself with your laptop connected to the Delco in real time (on the fly), or
- Contact your nearest Kalmaker Tuning Centre, so you can have your Delco ECM re-mapped professionally.

Kalmaker software comes in three configurations:
- Street Pro Series (suits ecm 1227808, 16183082, and 1227165). The latest version is the Street Pro 3.
- Workshop Series (suits all Australian ecm/pcm computer models excluding GEN 3/LS-1).
- Corvette/ Camaro/ Firebird Series (suits 1227165, the 727 & 730) GM TPI.

This section shall cover the Street Pro 3 configuration.

Street Pro 3

Street Pro 3 is a package that combines Kalmaker software, a real time ECU, cable to link the ECU to a laptop computer, the instruction manuals and CD and a security device called a Dongle. The software utilises many different features of GM Delco ECMs from vehicles such as the Pontiac Turbo and Buick Turbo. It suits manual or auto transmission, and the software contains lock up torque converter control to suit the TH-700 Automatic Transmission. It is not suitable for 4L60E electronic transmissions.[12]

---

12. Kalmaker's Workshop software can run electronic transmission. See Kalmaker on page 113

Street Pro 3 comes with many different configurations, stored in the software as Calibrations. The following procedure is a broad description of the tuning process:

1. Install the Kalmaker software onto your laptop.
2. Plug the dongle into the parallel port of your laptop.
3. Connect the vehicle to the real time ECM in the normal manner.
4. Connect the laptop to the real time ECM using the supplied cable.
5. Load one of the *Calibrations* that looks close to your specifications to get the engine started. It may be necessary to copy a calibration from a similarly modified engine, but close enough is good enough to get the engine started.
6. Before starting the engine, set injector size (Base Pulse Width).
7. Tell Kalmaker which sensors are installed (O2 Sensor, MAP, Knock Sensor, VSS, etc.).
8. Select *Auto Fuel Update*.
9. Start the engine.
10. Using the laptop computer, take the engine through its paces, from idle through cruise and then at WOT, making corrections to the *Variables* in accordance with the instructions.

Tuning the Engine Management System

"Tuning" the EMS is a bit misleading. The process of tuning involves very little hands-on mechanical interference, such as adjusting idle mixture screws, swapping power valves and jets or twisting the distributor around. In the case of engine management systems, tuning is the process of matching engine speed and load with fuel and spark. For a given load on the engine (determined by how much air is being sucked into it) the engine management system must pulse the injectors for the correct amount of fuel. Simultaneously, the EMS must fire the spark plugs at just the right time. Many other factors will alter the amount of fuel and spark, such as the amount of oxygen in the exhaust, altitude, air temperature, coolant temperature, etc. The way an EMS does this is quite amazing.

VE Table. Engine management software employs some form of VE (*Volumetric Efficiency*) table, which is a graph (or *map*) of engine speed vs engine load that starts and ends at two points - idle and WOT. At specified points along this line (for example, every 200rpm), Fuel *Injector Pulse Width* and *Ignition Advance* is determined. This linear torque curve is impossible to achieve, of course, so the VE table values are modified at each load point by data supplied by the sensors.

Base Pulse Width. For starters, you simply enter a list of known values, such as the number of injectors, injector size, cam duration, lift, engine displacement, compression ratio, transmission ratios, rear end ratio, tyre size, etc and the software calculates the essential data for you. The software may be embedded in a hand-held controller or installed on your laptop, but this initial exercise populates the appropriate tables with the data, and it's usually enough to get you cruising. This starting point is known as the *Base Idle* condition or the *Base Pulse Width*, referring to the length of time the injectors are open. Once this baseline has been established, it is time to "map" the fuel to engine speed and load. In other words, the engine requires more or less ignition advance and more or less fuel depending on engine speed and load. We use two components of our engine management system to do this:

- **Engine Speed (Crankshaft Position).** For SEFI systems, engine speed and piston position are determined by the Crankshaft Position Sensor (See *Crankshaft Position Sensor (CKS)* on page 61) and Camshaft Position Sensor (See *Camshaft Position Sensor (CPS)* on page 61). For Batch systems, only the Crankshaft Position Sensor is required (or, for even simpler systems, the TACH signal from a magnetic distributor or a MSD controller, such as the MSD 6AL).
- **Engine Load (MAP).** For Speed/Density systems, engine load is measured using the MAP (See *Manifold Absolute Pressure Sensor* on page 57). A MAF sensor assumes this role in mass air systems.

At the *Load Points*, you make adjustments to supply the correct amount of fuel, and then fire the spark with the correct amount of advance. The EMS software then looks for extra data supplied by the other sensors, which is stored in similar maps or *Lookup Tables* – so called because the software "looks up" the value. You already know what the data from the sensors *should* be – in fact, the EMS software may already have mapped the sensors with the most likely readings, so all you need to do is adjust the value slightly to get the best response from the engine. This is where a Dyno comes in handy, but it's not essential. If the software is doing its job, the engine will run close to stoichiometric (14.7:1) with "seat of the pants" tuning.

The data in each map is used to correct the base injector pulse width and base ignition advance until the engine is running at 14.7:1 A/F ratio. At each *Load Point*, you can make corrections that deliver the most power, the best economy and/or the best emissions – you can choose your priority. You lock your adjustments in by saving the data to the computer. You do this for every sensor at the specified number of *Load Points* (determined by the software). All the measured data is mapped to *Lookup Tables* for the individual sensors. Once you have tuned the EMS, the software does its job by constantly comparing the sensor reading with its associated data in the *Lookup Tables*. If the O2 sensor, MAP or MAF, Knock Sensor, MAT, CTS, etc. are all working properly, the engine will hum along nicely. If a sensor's input to the ECM does not match its value in the *Lookup Table* for the particular engine speed, timing and/or fuel is adjusted until it does. Remember, too, that all this happens thousands of times per second.

Fuel and Spark Compensations

Engines require adjustments to fuel and spark for a variety of reasons, such as cold starts, acceleration, deceleration, etc. The CTS (See *Coolant Temperature Sensor (CTS)* on page 54) provides compensations for engine temperature because you need an enriched fuel supply for a cold engine (we used to use a choke). The MAT (See *Manifold Air Temperature Sensor* on page 53) compensates for cold weather or hot weather, adjusting fuel enrichment accordingly. The IAC (See *Idle Air Control Valve* on page 59) provides fuel and air metering for the engine at idle. The O2 Sensor (See *Oxygen Sensor* on page 54) compensates for changes to the fuel:air ratio. It also compensates for idle speed when the Air Conditioning is switched on or when the engine is operating at high altitudes.

Using a Dyno makes it possible to squeeze every ounce of horsepower available from your engine, but this kind of fine tuning is usually unnecessary. It is also more convenient than making runs up and down the road.

By adjusting all the necessary variables, Kalmaker can be tuned specifically to your engine, taking into account all the modifications such as cam, compression, air-flow (exhaust/intake porting), ignition, gears (transmission, diff ratio and tyre size) to name a few. On the Real Time ECM, the engine data is updated accordingly. By employing Lean Cruise mode, you can tune for a nice, smooth idle, good economy at cruise and *grunt* at WOT. Wild cams that couldn't possibly idle under 1,000 rpm will now calmly rumble away at 600 rpm. Try THAT with carbs!

When the engine is humming nicely, you can switch everything off except the laptop and *burn* your engine's new configuration on to an Electrically Erasable Programmable Read Only Memory (EEPROM) chip and plug it into your ECM in place of the factory chip. Remove the Kalmaker Real Time ECM and put your newly configured ECM back in its place.

Alternatively, you can leave the Real Time ECM in place.

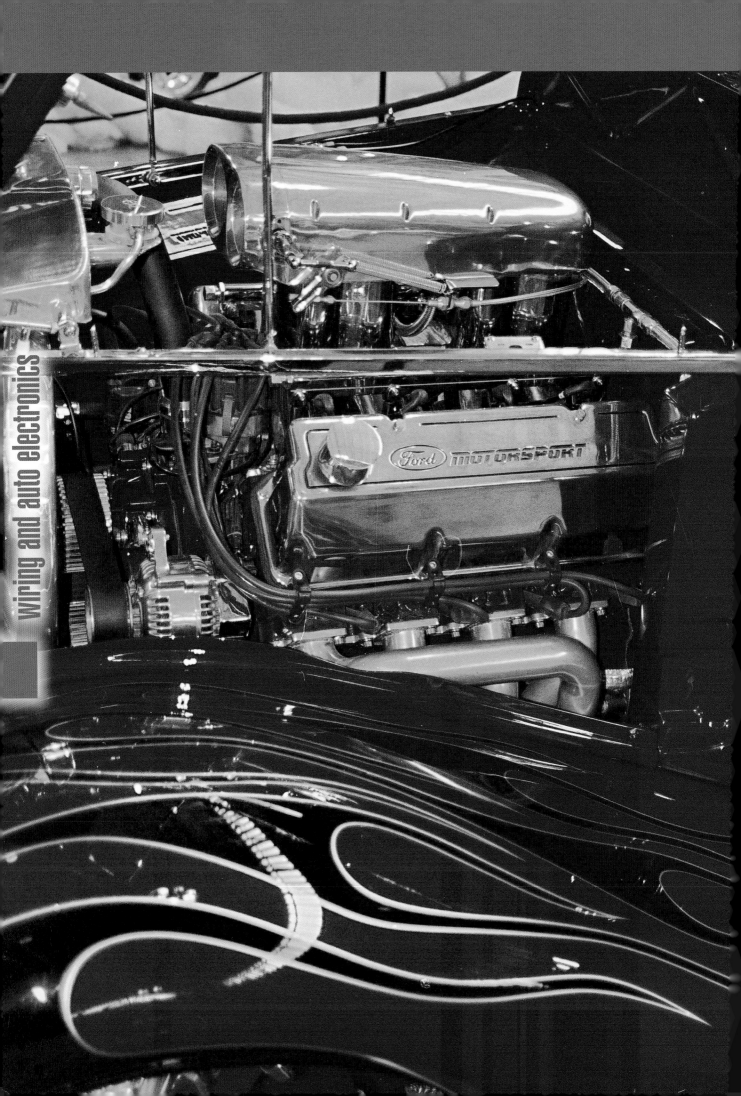

wiring and auto electronics

7 Wiring and auto electronics

Engine Management Systems include auto electronic components that must be treated differently to the rest of the vehicle's electrical system. Components such as the ECM, the Crankshaft Position Sensor/Ignition Module, Hall effect switches in the distributor, Camshaft Position Sensor or Vehicle Speed Sensor all require special attention because they can be easily damaged by incorrect or badly terminated wires. These components are also sensitive to electrical "spikes", such as those caused when the battery terminal is loose or some idiot is bouncing the battery supply lead against the battery terminal looking for sparks to see if something is drawing a load.

You can't do these kinds of things any more!

Other taboo practices include:
- Never disconnect the battery to test the alternator.
- Never jump start a car with unprotected jumper leads, and, even then, only if it's an emergency.
- Never reverse the polarity of the battery connections. You will damage the ECM, guaranteed!

> **Note:** This book is a Reference Manual. If you want information on how to wire up your Hot Rod, Custom or Street Machine project, I highly recommend Skip Readio's great book, 'How to Do Electrical Systems: Most Everything About Auto Electrics' on page 115.

The Alternator

A car's electrical system begins at the alternator. The battery simply maintains electrical continuity for the times when the alternator is not turning. Without going into lengthy technical theorems and explanations, we start at the alternator because the electronic devices "see" the alternator as the source of power. For engines that undergo conversion to EFI, selecting an appropriate alternator is important. Generators, remotely regulated early alternators, even the ones with "solid state" regulators are unsuitable because they do not provide sufficient shielding and current smoothing that later model alternators enjoy. At the very least, use a single wire alternator that normally came fitted to vehicles with electronic ignition. Preferably, use a late model type that comes from a vehicle that employs an engine management system.

FIGURE 40. High Output Alternators
In the not too distant future, alternator voltage will be around 42 volts instead of the 12 volts that we are used to. Modern alternators are shielded to minimise Electro Magnetic Fields (EMF) from interferring with electronic components. They are also driven by a serpentine belt and the output is regulated more efficiently, delivering a "smoother" DC voltage.

Remember, too, that a factory alternator is designed for a specific application in a specific vehicle. The speed at which it is designed to rotate, the voltage it will produce, the current limits, and other operating characteristics are built for the donor vehicle.

Alternator Rating

Engine Management Systems by themselves don't draw a great deal of current. The computer, sensors, relays, etc will not put a drain on the average charging system. Injectors will draw the most current, but, again, not beyond the capacity of your average system. It is important, however, that the rest of the electrical system in your project vehicle does not adversely affect electrical supply to the critical components, particularly the computer. Huge stereo systems, electric water pumps, air conditioning, compressors for air bags or hydraulic suspension systems all place large demands on the electrical supply that your alternator needs to maintain.

To determine your charging system's rating, check out the alternator itself. There may be a tally plate attached with all the details. If possible, look for the *Hot Rating* on the alternator. It may be a case of finding a co-operative auto parts store or car dealership and ask them to look up the alternator size/rating for the donor vehicle. The hot rating will tell you the amount of current the alternator can draw once the engine reaches normal operating temperature (this is a lower rating than the cold rating).

Note: Allow about 10 Amps as a safety buffer.

If the alternator rating is exceeded due to excessive load, then current will be drawn from the battery. If the alternator and battery combined cannot meet the demand, then the vehicle's electrical system cannot supply adequate power to the components - lights may dim, ignition may misfire and critical ECM supply voltages are compromised. If the current drain of your entire electrical system is going to be more than 10 Amps over the hot rating of the alternator, you should upgrade to a higher output alternator. For borderline cases, you can enhance the capacity of your system by replacing the existing battery with a larger one.

A quick note on Capacitors. Becoming more popular as a quick fix for "boom-boom" cars is the *Capacitor*. For huge stereo systems that dim the lights every time the sub-woofers hit that big bass note, a Capacitor (Cap) can be installed in the power supply to the amplifier to absorb the electrical spike. A Cap on its own DOES NOT create power - it stores battery voltage and discharges it quickly on demand. Charging the Cap drains electrical energy just like any other load. Unless you are installing a stereo system that will blow the doors off your car, a Cap is no solution to a poorly designed charging system.

Alternator Choice. If you use an alternator rated at 120 amps (max) and the total current draw from the electrical system (including the battery) is only 20 amps, the alternator only supplies the necessary current (20 amps) to maintain the voltage set by the voltage regulator. For more detail on alternator operation, refer to the *Custom Auto Electronics and Auto Electrical Reference Manual*, but just remember for now that the alternator monitors and regulates the electrical system's voltage. If the voltage starts to fall below the regulator voltage (approximately 13.8 Volts depending on the alternator), the alternator supplies more current to keep the voltage stable.

> ***Note:*** *A 120 amp alternator does not continuously produce 120 amps unless there is a sufficient demand on the electrical system.*

For your EMS conversion project, choose an alternator that will do the job based on the donor car and its factory electrical components. Add to the system any extra components that your project will require and factor in our 10 Amp safety zone.

Alternator Power Curve. Next, you should consider the vehicle's application and, therefore, the alternator's Power Curve. This is determined by the alternator's pulley ratio and idle speed.

> ***Note:*** *Electrical demand at idle speed for engines with electronic Engine Management Systems is critical.*

To avoid any problems, it is important to understand the alternator's capability at slow speeds. While it is true that the alternator output increases as it speeds up, the increase is NOT linear. That's why it's called a power *Curve*. This is very evident when using high output alternators - at idle, small changes in speed make big differences to output. Pulley sizes are determined for factory applications, which satisfies the factory Power Curve. For our custom application, however, things may not strictly adhere to the factory specifications. When using dress up pulleys, gilmer drives or serpentine drives, consider the pulley ratio. For street use, it should be at least 3:1. For low idle, automatic transmissions that will spend lots of time cruising, make the ratio closer to 3.5:1.

> ***Note:*** *Do NOT use low ratio pulleys for high output (over 120 Amps) alternators. Voltage drops off considerably under 2400 rotor RPM.*

Calculating Pulley Ratio and Rotor Speed. The alternator rotor RPM is not necessarily the same as engine RPM. To calculate the actual alternator RPM, determine the ratio between the two pulley diameters by measuring the diameter of the crankshaft pulley and the diameter of the alternator pulley.

Note: Measure V belt pulleys on total outside diameter. Measure serpentine belt pulleys across the top of the grooves the belt rides on.

Divide the diameter of the crankshaft pulley by the diameter of the alternator pulley.

Ratio = Crankshaft Pulley Diameter ÷ Alternator Pulley Diameter

Now that we know the ratio, we can now determine the rotor speed:

Rotor RPM = Pulley Ratio x Engine Speed

example; 2.1 x 870 = 1827 Rotor RPM

Voltage Gauge. Well, not really a gauge, but a meter. A Voltmeter. DO NOT use an Amp Gauge (Ammeter) in your EFI/EMS project. In the old days, generators required an ammeter because generators were current regulated, and the ammeter would have no more than 15 Amps running through it. Since the 70's, cars have used alternators, which are voltage regulated devices (in fact, many manufacturers, such as Dodge and Holden, labelled their gauge "Amps" when they were actually Voltmeters) so this is why we use Voltmeters and not Ammeters. Besides, the ammeter has to handle the entire electrical system's supply. That's just too big a task for a small device that sits in the dashboard with its terminals exposed, and the average alternator pumping 70 Amps through it.

Alternator Grounding. Finally, a word on grounding. The most important part of any custom built electrical system and a non-factory EMS is GROUNDING. If you ever have a problem, it's going to be GROUNDING!

First, make sure that the negative post on the battery has a *direct* connection to the chassis, the body, the engine AND the alternator. If the battery is in the back of the car, this is especially important. Buy some proper braided grounding strap and run the strap the length of the car, hitching it every 15cm (6 inches) to a clean part of the body/chassis. By clean, that means free of paint and rust - BARE METAL! Don't rely on the engine block to ground the alternator, run the ground directly to the casing or to the ground post, if fitted. Secure the studs or mounting points with clean, new bolts and use star washers directly between the surface of the chassis/body/engine. You can paint over the grounding points AFTER they have been secured.

Starter Motors

Enough has been written about starter motors, their operation and description, so for more information, see *Further Reading* on page 115. For your EFI/EMS project, however, a few things need to be mentioned about starter motors.

Old style starters draw a lot of current. They cause the voltage to drop considerably when they crank the engine over, and if this action is prolonged, the voltage to the ECM can be affected. In custom and non-factory installations, this is critical, and in trunk mounted batteries, problems are going to arise because of this. The answer is, of course, a Hi Torque starter motor.

These starters use a gear reduction method to turn the engine over. This means that the actual starter motor is half the size of the old style starters, and draw far less current. Because of their compact size, they can fit the most crowded of engine bays and their solenoids are virtually impervious to heat.

FIGURE 41.
Hi Torque Starter Motor

Wiring, RFI and EMI

Radio Frequency Interference (RFI) and Electro Magnetic Interference (EMI) are important things to consider when installing an engine management system. RFI not only affects your stereo system, but can also interfere with the operation of the computer. EMI, while a different form of interference, can have devastating affects on electronic components of your car. Both forms of interference can be treated in the same manner by using care when wiring up your project vehicle and by the placement and mounting of your ignition trigger system and ECM.

RFI is the spurious radiation of radio frequencies, which are harmonics of much lower frequencies like the firing of spark plugs. If your V8 engine is running at 3,000 rpm, the frequency of the ignition is 3,000 x 8 = 24,000 hertz (cycles per second) or 25Khz (kilohertz). This frequency is of little consequence, but it carries with it all the *Harmonics* of that frequency. A *Harmonic* is a multiple of that frequency, but as you get further away from 24Khz, the harmonic level decreases substantially (the technical term is *Intermodulation Products (IP)*). With today's powerful ignition systems, however, the level of dangerous harmonics can be high enough to cause problems, mostly minor, but easily prevented.

EMI is the interference caused by the rapid expansion and contraction of an Electromagnetic Field (EMF). Your alternator is constantly generating an EMF, but is easily suppressed with a simple capacitor (suppressor) at its output and a well-grounded casing. Your thermo fan, starter motor, heater motor, even the tailshaft, generate EMFs, but the EMI they produce is suppressed by the motor's casing which is grounded to the negative terminal of the battery. On fibreglass bodies, grounding is even more important than on steel bodied cars.

Some RFI/EMI is created by external forces that you have no control over, such as mobile phones, CBs, taxi radios, power lines, etc, and have been known to cause air bags to deploy or anti-lock braking systems to fail. RFI/EMI is easily suppressed, but if it is not treated properly it can cause engine misfires, sporadic and intermittent failure of sensors and/or modules and a horrible squawking on your stereo. Every project is different, but here are a few rules that you should follow when installing any electrical/electronic device in your car to suppress RFI/EMI:

- Mount the ECM away from the engine, preferably separated by the firewall. Look at where the factory cars have their ECM.

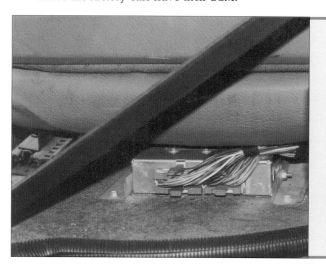

FIGURE42.
Computer Mounting
The computer in this Hot Rod has been mounted under the seat.
No problems here as long as the casing is securely grounded.

- Make sure the casing grounds are clean and tight, offering as much contact surface area as possible.
- Mount the ignition trigger box away from the ECM, also ensuring that it is well grounded.
- Never use solid core ignition wires.
- General wiring of electrical components should be of a high quality. I use and recommend Painless Wiring kits.

- Grounds, grounds, grounds. This cannot be overstressed.
- Shielding is the metal casing around electronic components such as stereos, digital gauges, the casing on the alternator, etc. Make sure they are all grounded.
- Use good quality ratchet-style crimping tools when making your wiring system.
- Use good quality spark plugs and don't let them wear out to the point that the electrodes start to pit and chip.
- Don't use solder on your wiring harness, it will crack and corrode at the joint.

Wiring Harness.
There are a lot more electrical connections to be made to the engine than in early carburetted engines. Injectors, sensors, trigger wheels and relays must all be considered when planning your wiring system. Factory engine harnesses are an easy option, as the distance between the connections and the computer will be fairly consistent. If you choose wisely, you can employ the same sensors and actuators in your project that are found on the donor vehicle, so the plugs and connectors on the engine harness will be consistent. Most programmable engine management systems come with a wiring harness and a standard set of sensors.

There are many aftermarket manufacturers of wiring harnesses and wiring kits for just about any project. Centech, Painless Wiring, Ron Francis, Classic Car Connections, Aussie Auto Looms and many others can supply all your general wiring needs. It is strongly recommended that you use one of these kits or use a factory engine wiring kit in its entirety rather than manufacture the engine wiring harness at home. This is because wiring of the ECM and sensors/actuators is critical, and home made systems just don't cut it.

Painless Wiring.
Arguably the leaders in their field, Painless Wiring's GM, Ford and Universal kits are based on GM and Ford factory connectors and colour codes. Unless your original wire harness is in great condition, and you have good diagrams telling you where each wire connects, it's easier just to replace it all, and it's quite likely that there's a Painless Wiring kit that will suit your needs. If you choose to utilise your factory harness, remember to take into consideration any non-factory accessories that may strain the current-carrying capacity of any single wire. For instance, electric cooling fans, electric water pumps, electric windows, door locks, trunk/rumble seat lifts, driving lights, hydraulic/air bag suspensions, air conditioning and electric fuel pumps, to name a few, will cause grief if your electrical wiring is not built to handle them.

For wire gauge and wire length calculations, refer *Wire Gauge Tables* on page 116.

Relays

If you add an electrical device, particularly one with an electric motor (for example, electric cooling fan, water pump or fuel pump) you should connect it to your wiring system using a relay. One relay should be used for each device.

The relay will handle the added current that the device would drain from your electrical system. You can activate a relay using a manually operated switch or some other switching device such as a thermo-activated switch, a vacuum activated switch or an RPM activated switch, common with Engine Management Systems.

There are two basic circuits in a relay:

- **Relay On/Off.** Current passes through two of the relay terminals to turn it ON. When current stops running, the relay turns OFF. In a common Single Pole Single Throw (SPST) relay, these are terminals 85 and 86.
- **Power.** Electrical power to run the device is connected through terminals 30 and 87.

In a vehicle that has been converted to EFI, it is not unusual to have several relays. For this

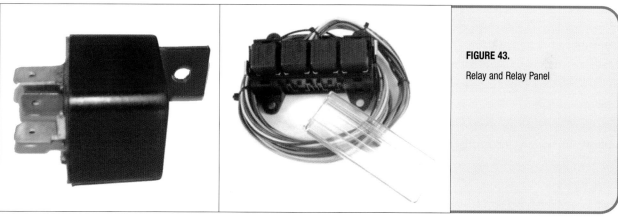

FIGURE 43.

Relay and Relay Panel

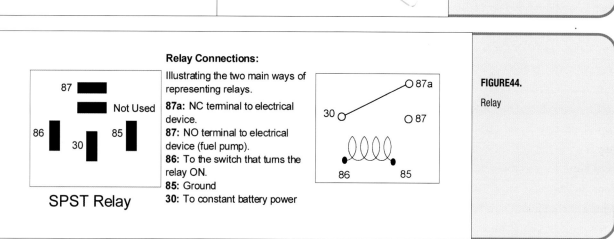

Relay Connections:

Illustrating the two main ways of representing relays.

87a: NC terminal to electrical device.
87: NO terminal to electrical device (fuel pump).
86: To the switch that turns the relay ON.
85: Ground
30: To constant battery power

FIGURE 44.

Relay

reason, a relay panel with sockets that accept the common SPST or SPDT (Single Pole Double Throw) relay makes life much easier.

Relay Jargon. SPST - Single Pole Single Throw means that there is one input (Single Pole) and one output (Single Throw). A SPDT - Single Pole Double Throw has one input and TWO outputs. The terms Normally Open and Normally Closed (NO, NC) are used to describe what the power contacts are doing at rest. For a SPDT relay, the usual configuration is one output is Normally Closed (NC, disconnects when the relay is switched ON) the other is Normally Open (NO, connects when the relay is switched ON). This means that the SPDT relay can control two devices at once.

Relay Rating. Relays are usually rated by the current handling capability of their power contacts. If it is rated at 40 amps, then pins 85 and 30 in the example illustrated in Relay in FIGURE 43 can handle up to 40 Amps of current. Check the current rating of the device you are installing and make sure the relay is rated accordingly.

> *Note:* *Relays should be mounted with the terminals down, otherwise they can collect water if the area is pressure washed or splashed.*

One of the most important relays you will need is the electric fuel pump relay to control when the pump is on and to ensure it turns off in case of the engine stalling - this is used to reduce the risk of fire (due to the pump still pumping fuel) in case of an accident and a ruptured fuel line. Look for a correctly rated SPDT relay and connect the fuel pump supply via the NO contacts. A good tip here is to use the fuel pump relay that came on the donor vehicle. Aftermarket in line fuel pumps usually come with the correct relay.

Tachometers

Factory and after-market Tachometers do not always operate in a consistent manner. Some tachos are voltage-triggered - they receive pulses from the negative (-) side of the ignition coil. Others (usually factory tachos) are current-triggered and are connected to the positive (+) side of the ignition coil or to the output of the ECM/Trigger Box.

Electronic Ignition systems often have very high voltages at the the primary side of the ignition coil (sometimes more than 500 Volts) rendering most voltage type tachos useless as well as some current-type tachos. Most trigger boxes, such as the MSD and Holley Annihilator ignition systems have a 12 volt "square wave" tachometer output feature to safely and properly operate most aftermarket, voltage type tachos.

Some tachometers require a different signal, so a device such as a tacho signal conditioner, adapter or amplifier may be required between the trigger box/ECM and the tacho.

wiring and auto electronics

converting to efi

8 Converting to efi

A Hot Rod must have that "look" before it is considered a true Hot Rod. While there are many opinions on what the "Hot Rod Look" actually is, one thing is certain – you can't get it from a factory floor!

Converting to EFI is, to some people, detracting from "The Look". Having said that, no-one can deny the classic good looks of the earlier GM TPI, especially if it's tarted up with some polishing and an aftermarket air filter. It looks pretty good in any rod, whether you are a "traditional" rodder or not.

FIGURE 45. GM TPI
This '32 Ford Tudor looked great with a detailed TPI in the engine bay. Even the most staunch traditionalist would agree that the GM 1985 onwards TPI is a work of art, especially after having received the hot rod treatment, as depicted in this image of Mal Lawler's much travelled rod. This particular engine was the inspiration behind the first Hot Rod Handbook.

The same goes for the Ford EEC 4 and 5 EFI systems, although they are a little less symmetrical in appearance. By using a little Hot Rod ingenuity, however, you can transform the Ford EFI in a manner that better suits the non factory application.

All this is okay, provided you are building a car with these powerplants in mind. More popular every day is the quad cam Toyota V8 and the Nissan V8, especially when the prices are lower than any GM/Ford cut-out!

Convert to EFI

But what about the rods that are already on the street, still giving their owners miles and miles of rodding fun, that want to convert to EFI? Sure, you can get a complete drivetrain,

FIGURE 46.

**Ford 5.0L Custom Plenum
by Marshall Perron**

This Ford 5.0L engine powers a classic 1965 Mustang convertible. The owner/builder, Marshall Perron, didn't like the look of the factory plenum, or upper manifold, so he built his own from polished aluminum and stainless steel.

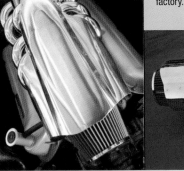

The custom conversion retains all the factory Ford 5.0L sensors and a K&N air filter takes over from the big black box provided by the factory.

FIGURE 47. Toyota Quad Cam V8

This Toyota Quad Cam V8 looks right at home in this early Studerbaker pickup. These engines are plenty powerful, cheap and abundant but you need the engine, trans, computer, wiring, maybe some changes to the cooling system, the exhaust, not to mention the dollars! Many factory engines have advanced to EFI while still retaining their basic engine block – the SBC and 5.0 Litre Ford come to mind. Chrysler commercial vehicles, too, have the 318/360 EFI systems that still retain their basic LA engine blocks. Get one of these manifolds and you're halfway to your conversion at half the cost. No need to drill the manifold and build fuel rails, and there's plenty of imaginative ways to get throttle bodies to look and work right. However, it's not always that easy, especially if you have one of those engines that doesn't have a factory EFI manifold at a reasonable price, or if the hi-tech look of the plenum and its associated plumbing is a put-off.

In the US, there are many shops that advertise manifold conversions and throttle body adaptations. Force Fuel Injection, BDS, Jim's Performance, Rance EFI, Kinsler, Hilborn, just to name a few. In Australia, Inner Active Manifolds in Blacktown, NSW, working closely with Ross Racing up the road, will accommodate just about any engineering feat. What's more, they manufacture their own brand of throttle body adapter, and supply the necessary hardware for the manifold conversion as well, all for a reasonable price.

Conversion Alternatives

Conversion from a carb/points car to EFI power using a factory EMS is okay if your donor engine stays stock as a rock. Modifying your engine means you must switch to a programmable EMS like Haltech or maybe you might choose to toss the factory EMS and go GM Delco 808 ECM tuned with Kalmaker. For most of us with existing 4-barrel manifolds and distributors, however, converting the manifold to EFI and using a factory or after-market throttle body is the solution.

If you are fortunate enough to come across a Hilborn or Enderle type stack injector that fits your engine, conversion from constant flow to EFI operation is the same process. Stack injection will give you the ultimate in "Hot Rod" induction with the look of a classic dragster and the street manners of EFI.

Notice that these options require some machining of the manifold. At the ends of the runners, holes need to be drilled so that injectors can be inserted. With the stack injection, constant flow injectors won't work, and they are much narrower than EFI injectors, so these manifolds need machining, too. Simple enough. And it is, if you know what you are doing, because if you know what you are doing then you know:

- The injectors must point at the back of the intake valve.
- The injectors must all be at the same depth.
- The injectors must all be at the same height so that the fuel rail, which joins them, is level.

FIGURE 48.

Stack Injection Alternative.

How about this for your Windsor, a stack injection set-up from EFI Hardware (Speed Technology) in Mitcham, Vic. They manufacture IDF/IDA Weber pattern throttle bodies for use on Weber manifolds such as the Windsor featured here. If you have a manifold, conversion to EFI is easy with these throttle bodies, and they look good too!

- The fuel rails must not leak (the fuel, remember, is at high pressure), and must be secure enough that the injectors don't shift in the event of a backfire.

Before you go any further, take your manifold to a reputable machine shop to convert it to EFI for you. For a cost of a couple of hundred dollars (check first, some manifolds may require more work, for example, tunnel rams) you should get:

- The holes milled and cut for equal depth.
- Injector bungs welded in to the holes.
- Inner runners ground smooth.
- Injector height set correctly by milling the injector bungs flat and square.
- Fuel rails cut and milled.
- Your injectors installed in the rails and secured into your manifold.

To find a machinist that can do this work, just contact any of the businesses featured under *Appendix* starting on page 109.

Choosing a Manifold

Before we look at the engineering side of things, here's some advice on manifold choice:
A single plane manifold (eg, Victor Jr., Torker, etc) works best, although the split plenums (eg, performer) will also work. A tunnel ram is ideal, but must be the type that has individual runners so that air is directed to the port. Boat type tunnel rams are completely open at the manifold base and flow way too much air, so stay away from them.

Aluminium manifolds are recommended for multi port injection conversion. That's because cast iron manifolds are too hard to machine and weld compared to the aluminium types. To retain your 4 barrel cast iron manifold, you would have to build a Throttle Body Injection (TBI) setup as opposed to multi point. There are a few TBI throttle bodies around that will bolt straight on to a 4 barrel manifold, such as those manufactured by EFI Hardware, Holley, Edelbrock, etc. You might also consider some of the two barrel Weber type TBI throttle bodies, or use a GM TBI adapter plate (See *Throttle Bodies and Adapters* on page 111) but these are "Wet Manifold" systems that we haven't got space for here. Suffice to say, it is a good alternative for some of the older and/or obscure engines, such as early Cadillacs, Nailhead Buicks, Ford Y-Blocks, etc where aluminium manifolds are either too scarce or don't exist.

Converting a Carb Manifold

Sticking a manifold in to a drill press at home and drilling holes in it will result in failure. A garage-built conversion would cost far more than necessary on things like fuel rails, injector bungs, and aluminium welding. Machine shops and engineering firms such as those described in Appendix on page 109 have all the measurements and settings for just about every type of alloy manifold there is.

Thanks to Ben Clothier of Ross Racing for the run down of the procedure.

Ben Clothier is setting up the milling machine to drill the holes in a Mopar hi-rise manifold. It's a two stage process - a hole is drilled, then milled to the correct size using a cutter.

The manifold is aligned on the bed to ensure the cut will be at the correct angle. The injector must point to the back of the valve.
The bit is fitted and holes in the runners are drilled to provide access for the cutting tool

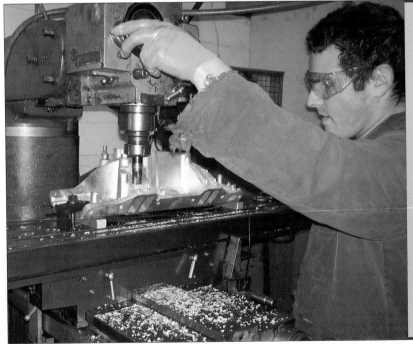

Now the cutter is used to shape the hole exactly so that the injector bosses, which are machined sections of thick wall extruded aluminium, can be inserted and welded in place.

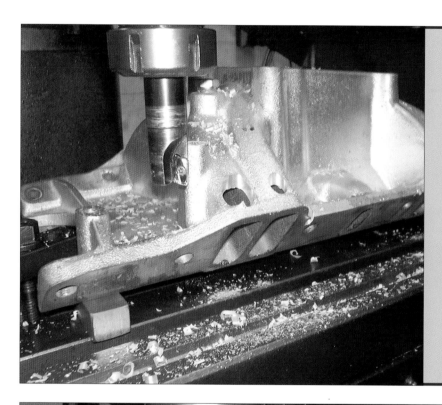

All done on this end. All four runners on one side have been drilled, then cut to size. The manifold is flipped around, re-aligned and the process begins again.

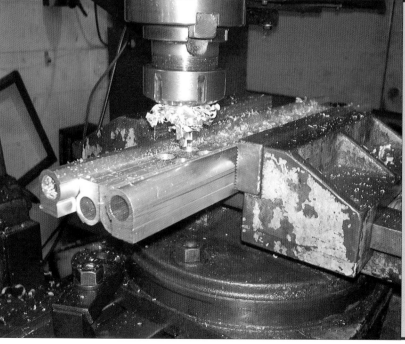

The manifold has been done, now we need to make the fuel rails. Again, this is a precise operation, and you need to know what you are doing. The fuel rail is supplied in bulk form, and is cut to size. The holes are drilled to the injector O ring size.

converting to efi

The manifold, ready for the injector bungs to be welded in. The first bung has been pressed in. Once they are all welded in, the tops of the bungs are milled flat and square to ensure they are all exactly the same height from the deck.

One down - seven to go!
Notice how close the two ports are on the Mopar manifold. Small block Chevys are the same, making it difficult for a novice welder to get a good seam.

Another example is this tunnel ram, all finished and waiting for the fuel rails. Tunnel rams pose more of a challenge for the machinist, as it is difficult to get the cutter at the correct angle.

Looking at the underside of the manifold, be sure to grind the injector bungs down so that they are flush with the runners. The illustration shows the bungs before grinding.

converting to efi

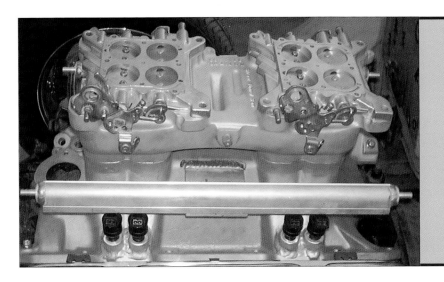

FIGURE 49. Throttle Bodies

Holley carb bases used as throttle bodies on this converted tunnel ram.

Using carb base plates as throttle bodies will work, but you need to adapt a Throttle Position Sensor. GM V6 and V8 TBI throttle bodies will also work, and the TPS and IAC are already mounted.

Adapting Throttle Bodies

One of the cheapest ways to get a throttle body for a four barrel manifold is to use a carb base. The drawback is the *Throttle Position Sensor* and the *Idle Air Control* valve – where do you mount them? See *Throttle Position Sensor* on page 58 and *Idle Air Control Valve* on page 59. You also need to adapt an air filter, but a couple of aluminium or phenolic spacers will do the job.

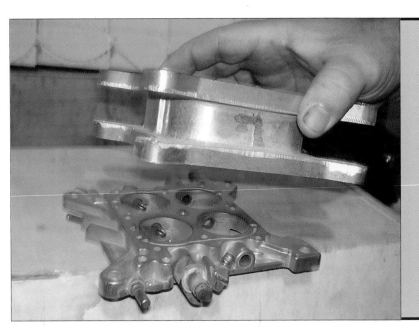

FIGURE 50. Providing Air Flow with Carb Spacers.

This Holley Carb base is what is being used in the tunnel ram above. By using a two inch aluminum carb spacer, available from most performance shops, an air filter or scoop can be mounted. The spacers also provide direction for the air flow, reducing turbulence even more when the edges are chamfered to a smooth curve.

Chev V6 and V8 TBI throttle bodies are plentiful and cheap, and you can adapt them to a four barrel manifold quite easily. The bonus is that factory TPS and IAC will fit without having to resort to fabrication of mounts.

An easier way is to use an after-market four-barrel throttle body, which has the TPS, IAC and air filter mount installed.

Another solution is to use two VN Commodore throttle bodies linked together on an adapter such as the Twin Tech adaptor illustrated (See *Throttle Bodies and Adaptors* on page 111). You can use the IAC and TPS from one of the throttle bodies (keep the others as spares). The Twin Tech is inexpensive and the Throttle Bodies are very cheap (they can be scrounged at wrecking yards or swap meets at bargain prices). New components are readily available from GM spares outlets.

For cramped engine bays, you might consider using the Ecotech throttle bodies, used in Australia on VS onwards Commodores. They are much shorter than the earlier VN throttle bodies, but retain the same mounting base, TPS and IAC.

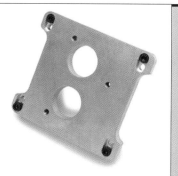

FIGURE 51. Three Bolt TBI Throttle Body Adaptor

Bolt on GM V6 or V8 throttle bodies to your four barrel carb manifold using these Painless Wiring Throttle Body Adaptors (Part Number 60118).

Convert from carburetion to TBI without having to replace your manifold. This adaptor bolts on any square or spread bore manifold and accepts any 43mm bore throttle body from 4.3L, 5.0L and 5.7L Chevy engines.

Using the TBI Throttle Body adaptors, twin throttle bodies can be used on this tunnel ram manifold. Airflow is difficult to control when using dual quads, but by limiting air flow with two smaller CFM V6 throttle bodies, correct air flow is maintained. This manifold is yet to receive Multi port treatment.

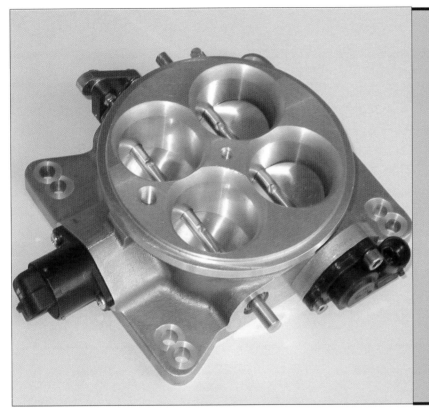

FIGURE 52.
Billet 4 bbl Throttle Body
from EFI Hardware

TPS, IAC and standard four barrel air filter are all accommodated in these Australian made four barrel throttle bodies.

They have adaptors for common GM TPS and IAC valves. They also have billet TBI throttle bodies.

Edelbrock and Holley also manufacture universal four barrel throttle bodies.

FIGURE 53. Twin Tech adaptor

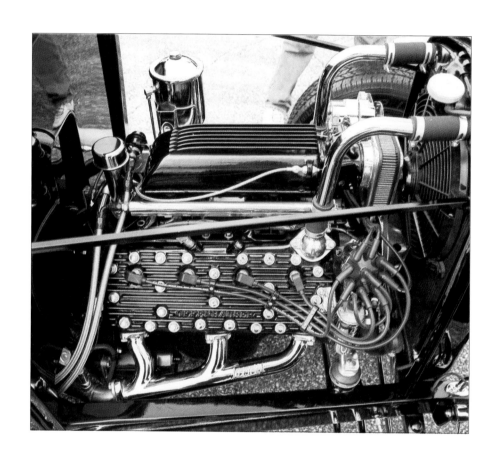

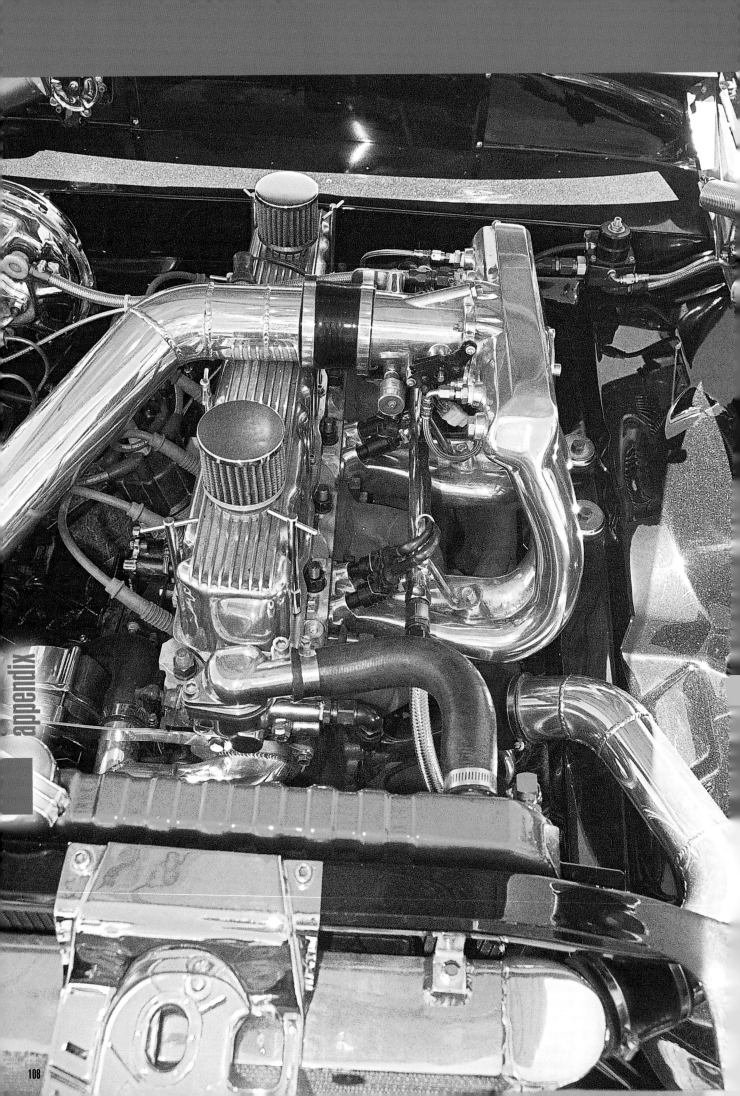

appendix 9

This section contains details of most of the commercial references used in this document. This is not an exhaustive list of all the manufacturers and suppliers of components and materials that are part of custom EFI and EMS conversions or projects. Instead, it is a collection of commercial enterprises that the author has come across while researching and fact finding for the content of this book. The author does not have a financial interest in any of the commercial enterprises mentioned. The reader must, therefore, take responsibility for their own actions if they make use of this information.

Fuel Rail Extrusion

There are three basic styles of fuel rail extrusion:

- *Dash 10.* Has a 0.8 inch bore.
- *Dash 6.* Has a 0.5 inch bore.
- *3/8 NPT.* Has a bore suitable for tapping to 3/8NPT.

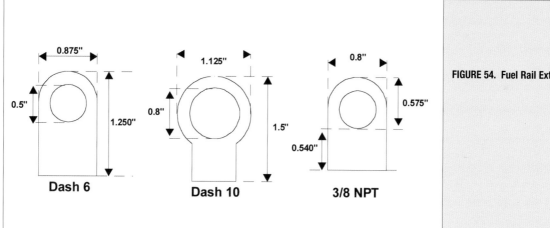

FIGURE 54. Fuel Rail Extrusion

The extrusion is usually constructed from high quality aircraft T6063 aluminium. The following information is provided for readers who are looking to purchase extruded aluminium fuel rail to make their own custom fuel rails. Prices (where indicated) should only be used as a guide. On line (Internet shopping) sellers are indicated by their URL.

> **Note:** These examples are **NOT** an exhaustive list. There are many more manufacturers and sellers of fuel rail, and the sizes can vary, depending on the application.

Ross Machine Racing. http://www.rossmachineracing.com/
The prices are $10.00 for -6 and $12.00 for the -10 per foot.

Canadian High Performance. http://www.hi-performance.com/index.html
Canadian Hi-Performance Discount sells rail in a number of lengths

Hi Octane Racing. http://www.hioctaneracing.com.au/Sard.htm
16 Euston St, Rydalmere, NSW Australia.

Arizona Speed and Marine. http://www.azspeed-marine.com/furaex.html
Phone: (480) 753-0208 Fax: (480) 753-0216

Electromotive Inc. http://www.electromotive-inc.com/efi_parts.html
Electromotive have dealers world wide.

O2 sensor bungs
O2 sensor bungs can be purchased from Castle Auto Electrics, Summit Racing (part number HLY-534-49), Painless Wiring (part number 60406) and most of the companies listed in this section.

Mopar Performance Electronic Ignitions
Mopar Performance has made it easy to buy most of what you need in a complete EI conversion kit. Kits are available for the following applications:

Table 3 - Mopar Performance EI

Application	Mopar Perf.Part No.
170-198-225 Slant Six	P3690789
3.9L V6	P4529402
273-318-340-360 Small Block V8	P3690426
361-383-400 B Engines	P3690427
413-426W-440 RB and 426 Hemi Engines	P3690428

If you are working with an existing system or performing your own conversion, many of the individual components are interchangeable. The pick-up coil is common to all applications, while the reluctor is different between 6 and 8 cylinder engines. Each of the applications listed in Mopar Performance EI above uses a different distributor. The stock trigger box is interchangeable with any application, but Mopar Performance markets several replacement units for high performance applications:

Table 4 - Mopar Performance Trigger Boxes

Application	Mopar Perf.Part No.
Orange Box for up to 6000 RPM	P4120505
Chrome Box for up to 8000 RPM (race)	P4120534
Gold Box for up to 10000 RPM (race)	P4120600

Calculations, Estimates and Equations

Here is an interactive web site that lets you put in fuel pressure, target Air:Fuel Ratios, cylinders and maximum duty cycle. It refers to a long list of Bosch injectors, discards the obviously inappropriate choices and displays for the remaining injectors the cc/hr and maximum HP.

http://www.bsmotor.com/?id=/cgi-bin/bsmotor/dyser/dyser.bsp?default

Throttle Bodies and Adapters

When converting from a Carb manifold to EFI, the Throttle Body needs to be compatible with the carb base bolt pattern. Modern factory throttle bodies are not designed for adapting to older carb manifolds. When modifying or transplanting an EFI engine, most factory throttle bodies won't supply the air, or just won't fit, so here are some suggestions which will solve these problems.

CFM-Tech

http://www.cfm-tech.com/catalog/

EFI Hardware in Mitcham, Vic, have a version which costs about the same as a new carb.
http://www.speed-technology.com

Inner Active Manifolds, manufacturers of the Twin Tech adapter.
http://www.manifolds.com.au

Holley, Edelbrock and Accel also manufacture four barrel throttle bodies and high performance factory replacement throttle bodies.

Painless Wiring have three bolt, Chevy V6/V8 TBI Throttle Body to four barrel square bore or spread bore adapters (Part Number 60118). 9505 Santa Paula Drive, Fort Worth, TX .
Phone 817-244-6898

http://www.painlesswiring.com

Remote IAC Valves

In the US, Fuel Air Spark Technology (FAST) and Full Throttle Performance can supply these units. Inner Active Manifolds in Blacktown offer a similar setup for their Twin Tech throttle body adapters. EFI Hardware also supplies a remote IAC. Injection Connection has several different remote IAC valves for any application as do Castle Auto Electrics.

EFI and EMS Businesses and Groups

Some of the aftermarket ECMs that are available are listed below:

SDS EM4

Racetech Inc.
1007-55 Ave. NE Calgary, Alberta, Canada T2E 6W1
Phone 403-274-0154, Fax 403-274-0556

INJEC Racing Development
137 Parer Rd Airport West, Vic, Australia 3042
Ph (02) 9338 4133 or 0414 338 413
http://www.sdsefi.com/

The EM4 Programmable Engine Management System from Simple Digital Systems in Canada or Injec Racing Developments in Victoria, Australia, will handle any engine you care to build, blown or unblown, while at the same time offering a simple programming solution in the form of a hand held controller. Contact details are available at the end of this article.

Haltech. *http://www.haltech.com.au/*
10 Bay Rd Taren Point NSW Australia, 2229
Ph: 61 2 9525 2400 Fax: 61 2 9525 2991
email: sales-au@haltech.com.au

MoTeC Pty Ltd *http://www.motec.com.au/*
Head Office 121 Merrindale Drive, Croydon South, Victoria 3136 Australia.
Phone: 03 9761 5050
Email: Technical enquiries: support@motec.com.au
General enquiries: admin@motec.com.au

Megasquirt *http://www.bgsoflex.com/megasquirt.html*
Build your own Engine Management System for $150. Includes a simulator, batch injection computer and software to set up your Megasquirt using a laptop. Excellent open system, lots of support from users around the world, used on all engines from Harleys to Hot Rods to dragsters. You will need some practice at soldering electronic components on to a printed circuit board and following some basic instructions. Then you can download the software from the web site and tune your engine with a lap top.

SST Automotive *http://www.sstautomotive.com.au/index.html*
Email: *sstwolf@iinet.net.au*

Here are some web sites that have a variety of products, product information and technical info.
http://www.fuelairspark.com/catalog/banktobank.asp
http://sandstallion.com/link.html
http://www.electromotive-inc.com/products/tec3.html
htttp://www.sdsefi.com/

Kalmaker

Alan Gibbs is the main distributor of Kalmaker, and his website is full of practical information: *http://www.kalmaker.com.au/*
Kalmaker is an Australian software company founded by Ken Young, a performance car enthusiast. Here is a list of some of the licensed Kalmaker engineers recommended by Alan and Ken:

> **Note:** Several Kalmaker kits, including Street Pro 3 and Workshop, have been sold in the US, but at the time of writing, none have confirmed that they are actively tuning computers. The Chev TPI software is now available for Kalmaker. Contact Alan Gibbs at Injection Connection for details and dealership enquiries.

NSW

Injection Perfection
Cnr Drover Rd & Desoutter St,
Bankstown Airport, Bankstown
02 9791 3122

A.L.S. Performance
2 Wattle St, Haberfield
02 9798 6511

Gossies Mechanical Services
9 Moira Park Rd, Morisset
02 4973 2128

Sams Performance
78 Gibson Ave, Padstow
02 9772 3105

Vic

Blue Chip Tuning
Deer Park
03 9266 2390   0411 404 659

Electronic Automotive
15 Salicki Ave, Epping
03 9401 5726

Amberley Auto
12-14 Amberley Cres, Dandenong
03 9793 1616

Chev Offroad
2 Station Rd, Caulfield
03 9571 9605

Jenkins Engineering
1 Gibson St, East Caulfield
03 8500 5764

SA

Awesome Automotive
12 Tobruk Ave, St Marys
08 8277 3927

QLD

ChipTorque
1/50 Lawrence Drv, Nerang
07 5596 4204

Paramount Performance
35 Wylie St, Toowoomba
07 4659 9711

Redcliffe Dyno & Performance
8 Cameron St, Clontarf
07 3284 1925

WA

Bill Lee Automotive
9 Ruse St, Osborne Park
08 9443 7741

Chipmaster
Samson St, Cannington
0412 770 636

Workshop: Unit 1
8 Samson St, Cannington
08 9452 0213

Injection Connection
14 Birch St, Attadale
08 9317 4750

GM Engine Management System ECM usage

All Australian 1227808 ECMs are MAP based and used in the following Australian vehicles

- 5.0L V8 VL Group A SS "Walkinshaw" Holden Commodore (1987, 500 built).
- 5.0L V8 VN and VP Holden Commodore (1991-1993).
- 3.8L V6 VN and VP Holden Commodore (1988-1991)
- 2.0L JE Holden Camira (1987-1992).
- 1.8L LD Holden Astra (1987-1989), and Nissan Pulsar (1987-1992).
- 1.8L JD Holden Camira (1986-1987).
- 1.6L LD Holden Astra (1986-1987).
- 1.6L Daewoo Cielo (1987).

The 1227165 ECMs are used on US '86-'89 V8 TPI (Corvette, Firebird, TransAm, etc), and some L4 TBI.

Further Reading

Small Block Chevrolet Tuned Port Injection
ISBN 0 949398 01 2

Auto Electronic and Electrical Reference Manual
ISBN 0 949398 73 X

How to Do Electrical Systems: Most Everything About Auto Electrics
by Skip Readio. ISBN 1878772066

Wire Gauge Tables

American Wire Gauge (AWG) sizes may be determined by measuring the diameter of the conductor (the bare wire) with the insulation removed.

Note: *Use thicker wire (lower gauge number) for borderline measurements.*

Table 5 - Metric to AWG Conversion Table

Metric Size mm2	AWG Size	Metric Size mm2	AWG Size
0.5	20	8.0	8
0.8	18	13.0	6
1.0	16	19.0	4
2.0	14	32.0	2
3.0	12	52.0	0
5.0	10		

Table 6 - Wire Gauge Diameter

American Wire gauge	Wire Diameter in inches	American Wire gauge	Wire Diameter in inches
20	0.03196118	5	0.18194
18	0.040303	4	0.20431
16	0.0508214	3	0.22942
14	0.064084	2	0.25763
12	0.08080810	1	0.2893
10	0.10189	0	0.32486
8	0.128496	00	0.3648
6	0.16202		

When choosing wire gauge, the distance the wire must run and the current (in Amps) it will be expected to carry must be determined first. In the following table, find the current (Circuit Amps) or the Power (Circuit Watts) the circuit is expected to carry. Now look up the gauge of the wire required for the length of the wire in the circuit.

Table 7 - Selecting a Wire Gauge

Circuit Amps Length	Circuit Watts (in Feet)	3'	5'	7'	10'	15'	20'	25'
0 to 5	30	18	18	18	18	18	18	18
6	36	18	18	18	18	18	18	16
7	42	18	18	18	18	18	18	16
8	48	18	18	18	18	18	16	16
10	60	18	18	18	18	16	16	16
11	66	18	18	18	18	16	16	14
12	72	18	18	18	18	16	16	14
15	90	18	18	18	18	14	14	12
18	108	18	18	16	16	14	14	12
20	120	18	18	16	16	14	12	10
22	132	18	18	16	16	12	12	10
24	144	18	18	16	16	12	12	10
30	180	18	16	16	14	10	10	10
40	240	18	16	14	12	10	10	8
50	300	16	14	12	12	10	10	8
100	600	12	12	10	10	6	6	4
150	900	10	10	8	8	4	4	2
200	1200	10	8	8	6	4	4	2

Oxygen Sensor Wire Colours

The following table lists the most common manufacturers of Oxygen Sensors fitted to GM, Ford, Chrysler and Mitsubishi vehicles, and the colour code of the wires so that you can use a factory O2 sensor in your project.

Table 8 - Oxygen Sensor Wire Colours

SENSOR MFG	1 WIRE	2 WIRE	3 WIRE	4 WIRE	FUNCTION
DELPHI	VIOLET	VIOLET	VIOLET	VIOLET	SIGNAL +
		PINK		PINK	SIGNAL -
			BROWN	BROWN	HEATER
			BROWN	BROWN	HEATER
DENSO	BLACK	BLUE	BLUE	BLUE	SIGNAL +
		WHITE		WHITE	SIGNAL -
			BLACK	BLACK	HEATER
			BLACK	BLACK	HEATER
BOSCH	BLACK		BLACK	BLACK	SIGNAL +
				GRAY	SIGNAL -
			WHITE	WHITE	HEATER
			WHITE	WHITE	HEATER
AUTOLITE	BLACK				SIGNAL +
NGK	BLACK			WHITE	SIGNAL+
				GREEN	SIGNAL-
				BLACK	HEATER
				BLACK	HEATER

appendix

Table 9 - Fuel Injectors

Flow (Lbs/Hr)	psi	Resistance	Colour	Style	Mfg	Mfg p/n
10.3	36.3				Bosch	280150207
12.8	36.3			MPI	Bosch	280150208
12.8	36.3				Bosch	280150210
12.9	43.5			MPI	Bosch	280150716
14		2.25	Dark Blue	MPI	Bosch	
14		2.35	Dark Blue	MPI	ND	
14	30	16.2	Gray	MPI	ND	
14	33	14.5	Gray	MPI	Bosch/ ND/DDK	280150727
14		11 to 18	Gray	MPI	Bosch	
14	39.2			MPI	Bosch	280150208
14.1	43.5				Bosch	280150211
14.1	43.5			MPI	Bosch	280150220
14.4	43.5			MPI	Bosch	280150703
14.4	43.5			MPI	Bosch	280150715
16.1	36.3			MPI	Bosch	280150204
16.1	36.3				Bosch	280150206
16.1	36.3			MPI	Bosch	280150209
16.1	36.3			MPI	Bosch	280150217
16.1	36.3			MPI	Bosch	280150219
16.4	36.3			MPI	Bosch	280150205
16.4	43.5				Bosch	280150704
16.4	36.3			MPI	Bosch	280150725
17	43.5				Bosch	280150209
17.2					Bosch	280150121
17.9	43.5				Bosch	280150100
17.9					Bosch	280150114
17.9					Bosch	280150116
17.9	43.5				Bosch	280150214
18.1	43		Gray	MPI	Bosch	280150105
18.2				MPI	Bosch	280150125
18.3	43.5				Bosch	280150614
18.3	43.5			MPI	Bosch	280150702
18.8	39.2			MPI	Bosch	280150203
18.8	43.5	15		MPI	Bosch	280150901
18.9	47.9			MPI	Bosch	280150775
19	40	2.25	White	MPI	Bosch	
19	32	16.2	Yellow-Orange	MPI	Bosch	280150718
19	33	16.2	Yellow-Orange	MPI	Bosch	
19	33	14.5	Yellow-Orange	MPI	Bosch / DDK	
19		11 to 18	Yellow-Orange	MPI		
19.3	36.3			MPI	Bosch	280150734
20.2	50	12.15		MPI	Rochester	17087325

Table 9 - Fuel Injectors (continued)

Flow (Lbs/Hr)	psi	Resistance	Colour	Style	Mfg	Mfg p/n
20.2	50	12.15		MPI	Rochester	17121947
20.7				MPI	Rochester	5235301
20.7				MPI	Rochester	5235302
20.7				MPI	Rochester	5235434
20.7				MPI	Rochester	5235435
20.7				MPI	Rochester	5235436
20.7				MPI	Rochester	5235437
20.7	36.3			MPI	Bosch	280150744
20.7	36.3			MPI	Bosch	280150157
20.7	36.3				Bosch	280150215
20.7	36.3				Bosch	280150216
20.7	36.3			MPI	Bosch	280150706
20.7	36.3			MPI	Bosch	280150712
20.7	43.5			MPI	Bosch	280150762
21.1	50	12.15		MPI	Rochester	17120254
21.1	50	12.15		MPI	Rochester	17121068
21.1	50	12.15		MPI	Rochester	17124248
21.5	36	16.15		MPI	Bosch	280150223
21.5	36	16.15		MPI	Bosch	280150239
21.8	39.2	14.5		MPI	Bosch	280150759
22	43.5	12.4		MPI	Rochester	5235357
22	43.5	12.4		MPI	Rochester	5235366
22	43.5	12.4		MPI	Rochester	17069647
22	43.5	12.4		MPI	Rochester	17069648
22	43.5	12.4		MPI	Rochester	17109952
22	43.5	12.4		MPI	Rochester	17109953
22	43.5	12.4		MPI	Rochester	5235451
22.2				MPI	Bosch	280150152
22.8	43.5		Dark Gray	MPI	Bosch	280150201
23		2.4	Black	MPI	Bosch	
23.2	29			MPI	Bosch	280150151
23.6	43					
24	43	14.5	Light Blue	MPI	Bosch	280150728
24	43.5			MPI		
24		11 to 18	Light Blue	MPI		
25						
25						
25						
25						
25.1	39.2				Bosch	280150422
25.6	43.5			MPI	Bosch	280150001
25.6	43.5			MPI	Bosch	280150002

appendix

Table 9 - Fuel Injectors (continued)

Flow (Lbs/Hr)	psi	Resistance	Colour	Style	Mfg	Mfg p/n
25.6	43.5		Yellow	MPI	Bosch	280150009
27	50	12.15		MPI	Rochester	17121909
27	50	12.15		MPI	Rochester	17124251
27	50	12.15		MPI	Rochester	17124289
27	55			MPI		
27	55			MPI		
27.4	43.5			MPI	Bosch	280150802
27.4	43.5			MPI	Bosch	280150802
28.7	55.1			MPI		
28.8	50.8			MPI	Bosch	280150811
29	43.5	HI Z		MPI	Bosch	280150151
29	43.5			MPI	Bosch	280150200
29	43.5			MPI	Bosch	280150335
29	43.5			MPI	Bosch	280150355
29	43.5			MPI	Bosch	280150357
29.8	45			MPI	Bosch	280150218
30	37	2.4	Green	MPI	Bosch	
30	37	2.35	Green	MPI	DDK	
30			Green	Barb	Bosch	280150024
30	43.5	11 to 18	Red	MPI	Bosch	280150756
30	43.5	11 to 18	Red	MPI	Bosch	280150911
30	55			MPI		
30				MPI	Rochester	17090844
30				MPI	Rochester	17121882
30.7	29			MPI	Bosch	280150035
31	43.5	11 to 18	Red	MPI	Bosch	280150912
31	43.5	11 to 18		MPI	Bosch	280150756
32	45	11 to 18	Red ?	MPI	Bosch	280150756
32	45	11 to 18		MPI	Rochester	15637677
32	55			MPI	Bosch	280150808
32.5	43.5			MPI	Bosch	280150804
32.8			Yellow	MPI	Bosch	280150007
32.8			Black	MPI	Bosch	280150044
32.8	39.2			MPI	Bosch	280155009
33.4	43.5			MPI	Bosch	280150951
35	43.5		Blue	MPI	Bosch	280150967
36.7	43.5			MPI	Bosch	280150003
36.7	43.5			MPI	Bosch	280150015
36.7	43.5			MPI	Bosch	280150024
36.7	43.5			MPI	Bosch	280150026
36.7	43.5			MPI	Bosch	280150043
37	40	2.0-2.25	Green	CFI	Bosch / DDK	280150402

Table 9 - Fuel Injectors (continued)

Flow (Lbs/Hr)	psi	Resistance	Colour	Style	Mfg	Mfg p/n
37.1	43.5				Bosch	280150814
37.7	39.2	HI Z	Green	MPI	Bosch	280150803
38.3	43.5			MPI	Bosch	280150835
38.6				MPI	Bosch	280150045
39.6				MPI	Bosch	280150035
41.9				MPI	Bosch	280410144
42.6	36.3			CFI	Bosch	280150608
46.1	33	2.00-2.40	Blue	CFI	Bosch / DDK	280150400
46.3	36.3	2.5		MPI	Bosch	280150036
46.3	43.5	2.5		MPI	Bosch	280150041
52.4	32	2.25	Gray	CFI	Bosch	
54.1	39.2	HI Z	Black	MPI	Bosch	280150351
55.6	16	1.4	Blue	CFI	Bosch	280150054
56						
57.9	36.3			MPI	Bosch	280410153
63.5	16	1.4	Green	CFI	Bosch	280150056
72		2		MPI	Rochester	17104988
77.2	72.5			MPI	Bosch	280410153
77.2	36.3			MPI	Bosch	280412911
80				MPI	Bosch	280411911
82	45	Low Z	Red		Bendix	25500139
108.1	72.5			MPI	Bosch	280412911

Flex Fuel & Methanol Injectors

25.3			Blue-Green	MPI		
38.3	43.5			MPI	Bosch	280150834
180	45	Low Z	White		Bendix	

Compressed Natural Gas Injector

N/A					Bosch	280150839
N/A					Bosch	280150837

appendix

Injector Flow Rate

The following table lists the maximum horsepower obtainable from injectors based on the injector flow rate operating on a 100% duty cycle. The horsepower figures assume a Base Specific Fuel Consumption (BSFC) of 0.55 lbs of fuel per hour at maximum HP (this is an average figure).

Table 10 - Injector Flow Rates

Horsepower @ 100% Duty Cycle

Injector Size Lb/Hr (cc/Min)	4Cyl	6Cyl	8Cyl
20 (210)	145	218	291
22 (231)	160	240	320
24 (252)	175	262	349
26 (273)	189	284	378
28 (294)	204	305	407
30 (314)	218	327	436
32 (335)	233	349	465
34 (356)	247	371	495
36 (377)	262	393	524
38 (398)	276	415	553
40 (419)	291	436	582
42 (440)	305	458	611
44 (461)	320	480	640
46 (482)	335	502	669
48 (503)	349	524	698
50 (524)	364	545	727
52 (545)	378	567	756
54 (566)	393	589	785
56 (587)	407	611	815
58 (608)	422	633	844
60 (629)	436	655	873
62 (650)	451	676	902
64 (671)	465	698	931
66 (692)	480	720	960
68 (713)	495	742	989
70 (734)	509	764	1018

Note: *You should choose injectors based on a duty cycle of no more than 85%.*

Note: *http://www.injector.com can build you any injector you want.*

Table 11 - List of Abbreviations

Abbreviation	Definition
A/F	Air fuel Ratio. Expressed in mass. 14.7 parts air to 1 part fuel is "stoich", the point at which, ideally, all of the fuel and all of the oxygen are consumed, with none left over. Less than 14.7 is rich, greater than 14.7 is lean
A/C	Air conditioning
ABDC	After Bottom Dead Centre
ABS	Anti-Lock Braking System
AC	Alternating Current
AE	Acceleration Enrichment. Similar to the accelerator pump on a carb. Additional fuelis injected for rapid throttle openings, or rapid decreases in MAP. Compensatesfor sensor lag, and also for fuel condensing out of the air when MAP decreases. Less AE is used on dry manifold (port injected) applications.
AFR	Air Fuel Ratio
AIR	Air Injection Reaction
AIT	Air Intake Temp
ALCL	Assembly line communications link. This is the serial line over which scan tools communicate with the GM ECM/PCM. Trouble codes are available, as well as various engine parameters.
ALDL	Assembly Line Diagnostic Link. See ALCL.
Async	Fuel delivery Asynchronous to engine timing
AT	Automatic Transmission
ATDC	After Top Dead Centre
AWG	American Wire Gauge
Baro	Barometric Pressure
BBDC	Before Bottom Dead Centre
BCM	Body Control Module
BLM	Block Learn Multiplier. Part of the self tuning capability of the ECM. An ECM typically has a table of BLM values, indexed by MAP and RPM. When the engine is in closed loop mode, the O2 sensor is monitored and fuel is either added or subtracted to maintain 14.7:1. Over time these changes accumulate in the BLM table.
BPW	Base Pulse Width
BTU	British Thermal Units
C	Celsius
C/L	Closed Loop
C3	GM ECM using Motorola 68xx-like processor (1+ MHz CPU)
CCP	Controlled Canister Purge
CEL	Check Engine Light (see MIL, SES)
CID	Cubic Inch Displacement
CKP	Crankshaft Position Sensor
CMP	Camshaft Position Sensor
CP	Canister Purge
CPU	Central Processor Unit (the brains in the ecm)
CTS	Coolant Temperature Sensor

Table 11 - List of Abbreviations (continued)

DB	xDecibels
DC	Duty Cycle On time, compared to total event time expressed as a percent
DC	Direct Current
DE	Decel Enleanment
DFCO	Deceleration Fuel Cut Off
DIS	Distributorless Ignition System
DVM	Digital Multi Meter
ECM	Engine Control Module
ECT	Engine Coolant Temperature
ECM	Engine Control Unit
EFI	Electronic Fuel Injection
EGR	Exhaust Gas Recirculation
EGT	Exhaust Gas Temperature
EPROM	Erasable Programmable Read Only Memory
ESC	Electronic Spark Control (Knock Control)
EST	Electronic Spark Timing
F	Fahrenheit
FED	Federal (emissions package)
GPS	Grams per second
HEGO	Heated exhaust gas oxygen
HEI	High Energy Ignition
HO2S	Heated o2 sensor
I/O	Input/Output
IAC	Idle Air Control
IAT	Intake Air Temperature
IC	Integrated Circuit
ICE	Internal Combustion engine
IGN	Ignition
IOT	Injector On Time. Time in milliseconds that an injector is on.
INT	Integrator
IPW	Injector Pulse Width
IS	Idle Speed
ISC	Idle Speed Control
kapwr	Keep alive power
kg	Kilogram
Kilohertz	(1,000 cycles / sec)
kPa	Kilopascals (Pressure)
KS	Knock Sensor
L	Litre (about 60 cubic inches)
LCD	Liquid Crystal Display
MAF	Mass Air Flow
MAP	Manifold Absolute Pressure
MAT	Manifold Air Temperature
MDP	Manifold differential Pressure

Table 11 - List of Abbreviations (continued)

MEMCAL	Memory Calibration Unit
MFI	Multi-port Injection
MIL	Malfunction Indicator Lamp (see CEL, SES)
MPFI	Multi Port, or Multi Point Fuel Injection, each cylinder has its own injector close to the intake valve
MT	Manual Transmission
NC	Normally Closed
NEU	Neutral
NO	Normally Open
NPTC	National Pipe thread Course
NTPF	National Pipe Thread Fine
O2	Sensor for detecting rich lean exhaust mixtures.
OBD2	On Board Diagnostics, newer vehicle electronics, US standard.
OL	Open Loop
OTS	Oil Temperature Sensor
P/N	Park Neutral Switch
P+H	Peak and Hold. High initial voltage to open then lower to keep it open
P4	GM ECM using early Motorola 68HCxx-like processor (2+ MHz CPU)
PCM	Powertrain Control Module (engine and trans control)
PE	Power Enrichment
PS	Power steering
PSI	Pounds per Square Inch
PSIA	PSI Absolute
PSIG	PSI Gauge
PW	Pulse Width
PWM	Pulse Width Modulation
QT	Quart
R-12	Refrigerant
RAM	Random Access Memory
RAP	Retained Accessory Power
RFI	Radio Frequency Interference
ROM	Read Only Memory
RR	Right Rear
RTV	Room Temperature Vulcanizing
RWD	Rear wheel drive
SEFI	Sequential Electronic Fuel Injection
SES	Service Engine Soon (see MIL, CEL)
SI	Saturated Injector *constant* voltage / current
SynchPulse	Fuel Pulse Width with delivery tied to timing of reference pulses
TAC	Thermostatic Air Cleaner
TB	Throttle Body
TBI	Throttle Body Injection system where the injectors are mounted above the butterflies

Table 11 - List of Abbreviations (continued)

TCC	Torque Converter Clutch
TCM	Transmission Control Module
TDC	Top Dead Centre
TFT	Trans Fluid Temperature
TH	Turbo-Hydramatic
TPI	Tuned Port Injection
TPS	Throttle Position Sensor
TRC	Throttle Return Control
TTS	Transmission Temperature Sensor
TV	Throttle Valve
TVS	Thermal Vacuum Switch
VATS	Vehicle Anti Theft System
VE	Volumetric Efficiency
VSS	Vehicle Speed Sensor
WBO2	Wide Band O2 sensor. This sensor differs from the O2 in so far that it is designed for a *linear* output
WG	Waste Gate
WOT	Wide Open Throttle

index

A

Actuators	33
Air Charge Temperature	53
Alpha-N	73
Alternator	
gilmer drives	86
Hot Rating	85
Power Curve	86
pulley ratio	86
rotor RPM	86
American Wire Gauge (AWG)	115
Amp Gauge (Ammeter)	86
Aquamist	75

B

Bank Injection	25, 71
Banks	25, 71
Base Idle	36, 58, 80
Base Idle Screw	36
Base Pulse Width	80
Boost cut	75

C

Camshaft Position Sensor	25
Capacitive Discharge	49
Capacitive Discharge (CDI)	44
Cast iron manifolds	98
Chip	72
Closed Loop Electronic WastegateControl	75
Closed Loop Fuel Control	75
Closed Loop Idle Control	75
Coarse Phasing	45
Cold Start Injector	36
Constant flow injection	27
Control notch	45
Crane Fireball	44

D

Delco	76
Deluxe Idle Control,	78
Direct Fire Ignition	42, 48
Disclaimer	14
Dongle	78
Dry Inlet Manifold	24

E

ECU	
damage to	15
EDIS Ignition Module	26, 48
Electrically Erasable Programmable	
Read Only Memory	81
Electromotive HPX	45
Electronic Control Module (ECM)	19
Electronic Control Unit (ECU)	19
Electronic Ignition	
Hall Effect Sensor	44
Magnetic Pickup	43
Reluctor Wheel	43
Electronic Lean Burn	43
EM1/EM2	112
Enderle	97
Environmental Issues	11

F

Factory calibrations	77
Fine Phasing	46
Flat Attack Racing	28
Ford Duraspark	43
Ford EDIS System	48
Ford EEC	74, 95
Fuel Injectors	
duty cycle	65
Flow Rate	65
High resistance	65
Low resistance	66

G

Generators	87
GM ignition module	49
GM TPI	74
Group Injection	71

H

Hilborn	97
Hot Rod Handbooks	11

I

Igniter	46
Ignition Advance	41
Ignition module	43
Ignition Retard	41
Injec Racing Developments	112
Inner Active Manifolds	67, 68
Integrated Circuit	72

J
Jump starting	15
Jumper Leads	
reverse-current protected	15

K
Kalmaker	113
Ken Young	76
Knock Sensor Control	75

L
Lean Cruise	78
Limp Back Mode	33
Limp Home Mode	33, 74
Load Point	80

M
MAP sensor	79
Alpha-N	36
Mass Air	79
MegaSquirt – Spark	26
Megasquirt – EDIS	26
Mike Davidson	27
MSD Flying Magnet	46

N
Not Enough Advance	42

O
O2 Sensor Mounting	58
Open Loop	38
Oxygen Sensor	
air/fuel ratio	57
Open Loop	32
voltage	58

P
Peak and Hold injectors	70
Powertrain Control Module (PCM)	19

R
Real Time Diagnostic	79
Real Time Editing	79
Reference Angle	45
Rev limiter	80
Ross Racing	106
Rotor Phasing	45

S
Self Learning	36, 83
Self Tuning	79
Sensors	33
Sequential Electronic Fuel Injection	70
Service Engine Soon	83
Single Fire	79
Single Pole Single Throw (SPST)	98
Spark Correction	80
SPDT	99
Speed Alpha	36
Speed Density	57
Advantages	36
Spiral core ignition cables	18
SPST	99
Stack Injection	71
Alpha-N	35
Stoichiometric	32

T
Throttle plate stop screw	38
Too Much Advance	42
Top Dead Centre	42
TPS Mode	36, 79
Trigger Boxes	45
Trouble codes	83
Twin Tables	79
Twin Tech	62

V
Variable Reluctance	50
VE Fuel Update	79
Voltmeter	93
Volumetric Efficiency	85

W
Wasted Spark	48
Water Injection	80
Welding	
danger to ECU	17
Wet Manifold	106
Wide Band Oxygen Sensor	37